PSpice for Windows
Volume II

Operational Amplifiers and Digital Circuits

Roy W. Goody

Mission College, Santa Clara, CA

Prentice Hall

Englewood Cliffs, New Jersey   Columbus, Ohio

Library of Congress Cataloging-in-Publication Data

Goody, Roy W.
    PSpice for Windows : operational amplifiers and digital circuits, vol. 2
        p. cm.
    Includes index.
    ISBN 0-13-235979-0
    1. PSpice for Windows. 2. Electric circuits—Computer simulation.
3. Electronic circuits—Computer simulation. I. Title
TK454.G66 1996   94-34322
621.3815'01'1353—dc20   CIP

Editor: Dave Garza
Production Editor: Christine M. Harrington
Production Manager: Pamela D. Bennett

 © 1996 by Prentice-Hall, Inc.
A Simon & Schuster Company
Englewood Cliffs, New Jersey 07632

All rights reserved. No part of this book may be reproduced, in any form or by any means, without permission in writing from the publisher.

Printed in the United States of America

10 9 8 7 6 5 4 3 2 1

ISBN: 0-13-235979-0

Prentice-Hall International (UK) Limited, *London*
Prentice-Hall of Australia Pty. Limited, *Sydney*
Prentice-Hall of Canada, Inc., *Toronto*
Prentice-Hall Hispanoamericana, S. A., *Mexico*
Prentice-Hall of India Private Limited, *New Delhi*
Prentice-Hall of Japan, Inc., *Tokyo*
Simon & Schuster Asia Pte. Ltd., *Singapore*
Editora Prentice-Hall do Brasil, Ltda., *Rio de Janeiro*

Contents

| Preface | v |

Part I — Feedback

Chapter 1	The Differential Amplifier—Common-mode Rejection	3
Chapter 2	Negative Feedback—Open Loop versus Closed Loop	11
Chapter 3	Subcircuits—Creating Symbols	17
Chapter 4	The Operational Amplifier—Specifications	25
Chapter 5	The Non-Inverting Configuration—VCVS & VCIS	33
Chapter 6	The Inverting Configuration—ICVS & ICIS	47
Chapter 7	Op Amp Integrator/Differentiator—Calculus	59
Chapter 8	Oscillators—Positive Feedback	71
Chapter 9	Filters—Passive and Active	81
Chapter 10	The Instrumentation Amplifier—Precision CMRR	89

Part II — Digital

Chapter 11	TTL Logic Gates—Analog/Digital Interfaces	97
Chapter 12	Digital Stimulus—File Stimulus	115
Chapter 13	Flip-Flops—Edge versus Level Triggered	129
Chapter 14	Digital Worst Case Analysis—Ambiguity and Timing Violations	141
Chapter 15	Critical Hazards—Persistence	157
Chapter 16	Counters—Synchronous versus Asynchronous	165
Chapter 17	Digital Coding—Comparators, Coders & Multiplexers	177
Chapter 18	The 555 Timer—Multivibrator Operation	185

PSpice for Windows

Part III — Filter Synthesis

Chapter 19	Passive Filters—Biquads	**195**
Chapter 20	Active Filters—The Switched-Capacitor Filter	**205**

Part IV — Applications

Chapter 21	Touch-Tone Decoding—The Bandpass Filter	**213**
Chapter 22	Pulse-Code Modulation—Time-Division Multiplexing	**223**
Chapter 23	Serial Communication—RS-232C	**233**
Chapter 24	Big-Ben Chime Control—The Comparator	**241**
Chapter 25	Alarm Entry—The Encoder	**247**

Appendices

Appendix A	Notes	**253**
Appendix B	Probe's Mathematical Operators	**254**
Appendix C	Scale Suffixes	**255**
Appendix D	Spec Sheets	**256**

Index		**265**

PSpice for Windows

Preface

What's past is prologue
William Shakespeare

This text is a continuation of *PSpice for Windows, Volume I,* also published by Prentice Hall and written by this author. *This advanced text assumes that you have acquired a working knowledge of the PSpice techniques presented by Volume I.*

If a prologue is a course of action foreshadowing greater events, then *Volume I* is truly a prologue to *Volume II*. Take all that you have learned in *Volume I,* covering DC/AC, devices & circuits, and the fundamental PSpice techniques, and add to that the feedback, digital, filter design, and advanced techniques of this text. Mix all this together, blend in experience, skill, and artistry—and step into an amazing world of circuit design where the only limitation is your imagination.

ORGANIZATION

The first ten chapters of Part I (*Feedback*) emphasize the operational amplifier in the closed loop configuration. The eight chapters of Part II (*Digital Circuits*) cover all aspects of basic digital, from the simple gates to synchronous counters. The two chapters of Part III (*Filter Synthesis*) cover advanced filter design using special filter synthesis software, and the final five chapters of Part IV (*Applications*) make use of all the previous material to design and test more complex analog/digital projects.

PREREQUISITES

Besides the ability to perform simple mathematical operations, the only prerequisite needed is a basic understanding of the theory and techniques presented in Volume I.

PSpice for Windows

A SUGGESTION

As with Volume I, we recommend that students also receive "hands-on" experience by prototyping actual circuits and troubleshooting with conventional instruments.

An approach that is useful when computers are limited is to divide students into two or more groups and rotate between PSpice and "hands-on." It is especially instructive to perform the same activity using both PSpice and hands-on, and to compare the results. *In this regard, most of the experimental activities outlined in this text can be performed using either PSpice or "hands-on."*

THE EVALUATION VERSION

This text is based on PSpice evaluation version 6.1. *Newer versions of the PSpice software are constantly being released and the chances are good that they will work with this manual. In general, you should use the latest version that is available; if any adjustments are necessary, they should be minor.*

Remember that MicroSim Corporation has made available these *evaluation* disks at no cost, and copying is "welcomed and encouraged."

GETTING STARTED

The software installation, mouse conventions, and circuit analysis processes are nearly identical with those presented in Volume I. Small differences in installation exist between the various versions (such as the pre-installation of Win32 software for versions 6.1 and later).

Because a 6.1 version of the *filter design* software was not released, we use version 5.4 for the Part III filter design chapters. Therefore, in order to perform the two filter design chapters, you may have to write MicroSim and request a copy of their special filter design software. For versions 6.1 or earlier, order software *fseval54*, and run file *filter.exe*.

If you encounter any problems during installation or use, refer to *PSpice for Windows, Volume I*, any MicroSim literature that might be available, the help files that come with the software, or contact MicroSim technical support at (714) 837-0790.

PSpice for Windows

CREDITS

As with Volume I, I would like to give special thanks to Debbie Horvitz of MicroSim Corporation for her natural friendliness and geniality, and for her many helpful suggestions and comments. Of course, MicroSim Corporation deserves special credit for making the evaluation disk available at no cost.

Many thanks again to Copy Editor Lorretta Palagi for her amazing ability to be both quick and thorough, and to make many useful suggestions concerning content and format.

Finally, I wish to express my sincere gratitude to Production Editor Christine Harrington and Administrative Editor David Garza of Prentice Hall Publishing. Under their professional guidance and management, the project moved smoothly forward and was released right on time.

Thank you for adopting *PSpice for Windows Volume II*; good luck and good success.

Roy W. Goody
Mission College

PART I
Feedback

In Part I, we concentrate on the science of *cybernetics*, where the power of *feedback* gives conventional circuits amazing properties. We will find that the *differential amplifier* is the ideal circuit to implement feedback.

A precision differential amplifier is known as an *operational amplifier*, and is one of the most powerful analog devices available to the designer. When the operational amplifier is combined with negative and positive feedback, we create such devices as amplifiers, buffers, integrators, differentiators, oscillators, and filters.

CHAPTER 1

The Differential Amplifier
Common-mode Rejection

OBJECTIVES

- To analyze the differential amplifier.
- To compare the differential mode with the common mode.

DISCUSSION

A differential amplifier has two inputs. It strongly amplifies the *difference* between the two inputs, but rejects signals that are *common* to both inputs. Such an amplifier is shown in Figure 1.1.

- Signals that are common to both inputs are forced to pass through the large 5kΩ emitter bias resistor and are rejected.
- Signals that are differential to both inputs pass easily through two small re' values and are amplified.

As an added bonus, the differential amplifier of Figure 1.1 requires no coupling or bypass capacitors and is therefore a *DC amplifier* (one with no low-frequency roll-off). The split power supply eliminates the input coupling capacitor by placing the input leads at ground, and the symmetric arrangement of the transistor pair eliminates the bypass capacitor because each transistor bypasses the other.

As we will see in the following chapters, the differential amplifier provides a far more important function than simply amplifying differential-mode DC signals: it provides a convenient method of implementing the "magic" of *negative or positive feedback*.

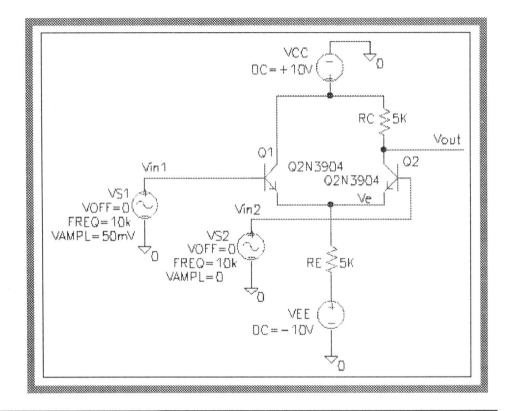

FIGURE 1.1

The differential amplifier

SIMULATION PRACTICE

1. Draw the circuit of Figure 1.1 and set the attributes as shown. (Note that VS2 presently acts as a ground because its amplitude is zero.)

2. Using both hand calculations and PSpice (bias point analysis), determine the Q point of transistor Q2 and enter your answers into Table 1.1. (Assume DC/AC *Beta* equals 175.)

    Do theory and experiment approximately match?

    Yes   **No**

    Does the bias current seem to evenly split between the two transistors?

    Yes   **No**

PSpice for Windows

	Calculations	PSpice
I_{CQ}	_____	_____
V_{CEQ}	_____	_____

TABLE 1.1
Bias point results

Differential mode

3. Using hand calculations, determine A_{DM} (differential mode gain), Z_{IN}, and Z_{OUT}, and place your answers in the "Calculations" column in Table 1.2

	Calculations	PSpice (transient)
A_{DM}	_____	_____
Z_{IN}	_____	_____
Z_{OUT}	_____	_____

TABLE 1.2
Differential mode results

4. Using PSpice (transient mode), determine A_{DM}, Z_{IN}, and Z_{OUT}, and place your answers in the "PSpice (transient)" column in Table 1.2.

> NOTES
> - When using transient analysis, trace variables must be plotted *separately* in order to determine the peak values.
> - When using transient analysis, use peak-to-peak/2 to factor out any DC value and average the + and - peaks.
> - V_{OUT}/V_{IN1} is automatically *differential mode* gain because V_{IN2}'s amplitude is zero.
> - See Chapter 14, Volume I, for Z_{IN} and Z_{OUT} measurement techniques.

PSpice for Windows

5. Reviewing Table 1.2, are the calculated and PSpice results approximately the same?

   Yes   **No**

6. Based on the phase relationship between input and output, circle the appropriate label for V_{IN1}.

   Inverting   *Non-inverting*

7. Switch the amplitudes of V_{S1} and V_{S2} ($V_{S1} = 0$ and $V_{S2} = 50mV$). Again measure the phase relationship between input and output and circle the appropriate label for V_{IN2}.

   Inverting   *Non-inverting*

Common mode

8. Connect the circuit in the *common mode*, as shown in Figure 1.2. (Note the greatly increased value for Vin.)

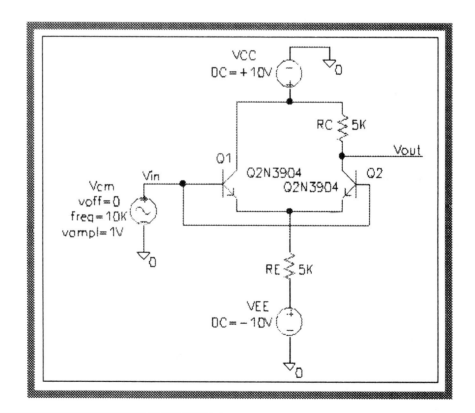

FIGURE 1.2

Common mode input

9. Using transient mode, measure the common-mode voltage gain and enter the result below.

   $A_{CM} =$ _____

   Comparing A_{DM} to A_{CM}, does the differential amplifier tend to reject common-mode input signals?

   Yes   No

10. As a measure of how much better differential-mode signals are amplified over common-mode signals, determine the *common-mode rejection ratio* (CMRR).

    $$CMRR = \frac{A_{DM}}{A_{CM}} = \underline{\qquad}$$

Advanced activities

11. Repeat steps 4-10 using the AC sweep mode. (Determine midband A_{DM} and A_{CM}, Z_{IN}, Z_{OUT}, phase, and CMRR.) Do the results generally agree with the calculated and transient mode results?

    > **NOTES**
    > - When using AC analysis, trace variables may be plotted directly in *combinational* form (such as V_{OUT}/V_{IN1}). (AC trace variables automatically display polar-form peak magnitudes by default, and all bias point [DC] components are automatically rejected.)
    > - Be sure to set the proper *AC=* attribute values to both VSINs.
    > - In the AC Sweep mode, use *P(T1/T2)* when determining phase <u>difference</u>, where T1 and T2 are any trace variables.
    > - See chapter 15, Volume I for Z_{IN} and Z_{OUT} measurement techniques.

12. While in the AC Sweep mode, determine bandwidth.

    BW = _____

13. Based on the values of A_{DM} and A_{CM} determined earlier, sketch a transient graph of V_{OUT} for the circuit of Figure 1.3. Compare your answer with the measured (PSpice) results.

14. The circuit of Figure 1.4 uses *current-source* biasing to improve CMRR. Measure the circuit's CMRR and compare your results to those of step 10.

PSpice for Windows

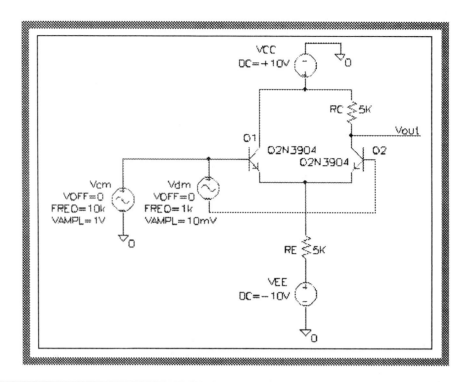

FIGURE 1.3

Combined circuit

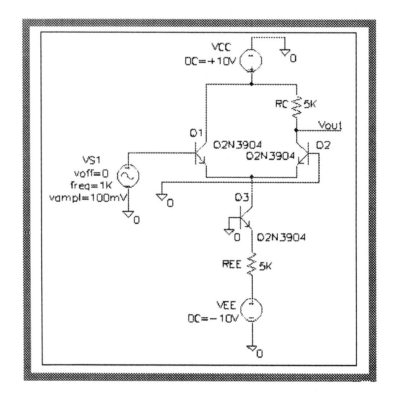

FIGURE 1.4

Using current-source biasing

Chapter 1   The Differential Amplifier

EXERCISES

- Design a differential amplifier that gives a differential (balanced) *output*. (Hint: add a 5k resistor to the collector of Q1 in Figure 1.4 and tap the voltage off both Q1 and Q2.) How does the gain compare to that of Figure 1.4?

QUESTIONS & PROBLEMS

1. Why is a *bypass* capacitor not required in the differential amplifier?

2. Based on Figures 1.1 and 1.2, write the formula for each of the following:

    (a) Differential-mode gain

    (b) Common-mode gain

3. How could you swamp the circuit of Figure 1.1 in order to reduce the harmonic distortion? What would happen to the voltage gain (A_{DM})?

4. When the circuit was modified as in Figure 1.4, why did the CMRR go up?

PSpice for Windows

5. Given the following, determine V_{OUT} as accurately as you can. (Note: assume that V_{S1} and V_{S2} are the same frequency and phase.)

   V_{S1} = 200mV, V_{S2} = 215mV, A_{DM} = 100, A_{CM} = 1

   V_{OUT} = _____

CHAPTER 2

Negative Feedback
Open Loop versus Closed Loop

OBJECTIVES

- To design a three-stage amplifier.
- To compare the properties of an open-loop and closed-loop amplifier.

DISCUSSION

In this chapter, we show how a magical "drop" of negative feedback can greatly enhance the properties of an amplifier. We start with the conventional three-stage amplifier of Figure 2.1—an amplifier that presently does not employ feedback (the *open loop* configuration).

Then, for just 5 cents in added components (two resistors), we add negative feedback (*close the loop*) and observe the amazing improvement in ease-of-design, stability, distortion, bandwidth, Z_{IN} and Z_{OUT}. We choose the differential amplifier of Chapter 1 for the first stage because it is ideally suited to carry out negative feedback.

Looking to Figure 2.1, we have a differential input state, a common-emitter middle stage, and a Class B buffer final stage. We note the complete absence of coupling and bypass capacitors, making the amplifier a DC amplifier.

- Coupling capacitors are not required because of the split power supply, and because the middle stage is also a *level-shifter* that allows for *direct coupling* between stages.
- Bypass capacitors are not required because the first stage is a differential amplifier, the second stage is a level-shifter rather than a high-gain amplifier, and the third stage is a buffer.

PSpice for Windows

SIMULATION PRACTICE

Open loop (without feedback)

1. Draw the three-stage open-loop (OL) amplifier of Figure 2.1 and set the attributes as shown. (We tap the first stage signal off Q1 to compensate for the phase shift introduced by Q3.)

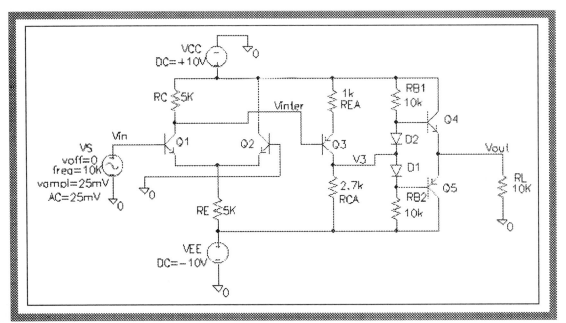

FIGURE 2.1
Three-stage amplifier
(open loop)

2. Using PSpice (transient, Fourier, and AC modes as needed), determine the following open-loop [OL] values: (<u>Note</u>: When using AC Sweep mode, all values are midband.)

   (a) A = _____   AdB = _____

   (b) BW (Bandwidth) = _____

   (c) HD (%total harmonic distortion) = _____

   (d) Z_{IN} = _____

   (e) Z_{OUT} = _____

PSpice for Windows

Closed loop (with feedback)

3. Add a negative feedback path to the circuit of Figure 2.1, and create the closed-loop (CL) amplifier of Figure 2.2. (Note the increase in V_{IN} from 25mV to 400mV.)

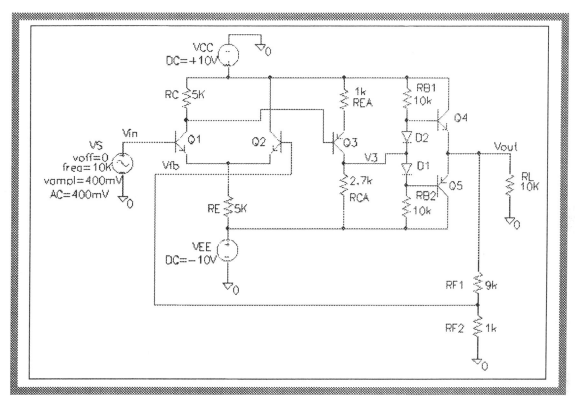

FIGURE 2.2
Three-stage amplifier (closed loop)

4. Using PSpice, determine each of the following closed-loop (CL) values:

    (a) A   = _____   AdB = _____

    (b) BW = _____   (3dB down to the right of the peak.)

    (c) HD = _____

    (d) Z_{IN} = _____

    (e) Z_{OUT} = _____

PSpice for Windows

5. As a final summary and comparison, enter the specifications of each type circuit side-by-side below:

	Open loop (OL)	Closed loop (CL)
A(reg) =	_____	_____
BW =	_____	_____
HD =	_____	_____
Z_{IN} =	_____	_____
Z_{OUT} =	_____	_____

6. Based on the results of Step 5, what conclusions can you draw about the effects of negative feedback?

Advanced activities

7. Perform a noise analysis of both open-loop and closed-loop amplifiers of this experiment and summarize your results. (Does negative feedback reduce noise?)

8. Reduce the load (RL) of both amplifiers (with and without feedback) from 10k to 1k and explain the results.

9. Set 10% device tolerances to all the resistors of the feedback amplifier of Figure 2.2, *except* the two feedback resistors (RF1 and RF2) Run a worst case analysis and note the results. Then reverse the configuration (only RF1 and RF2 have 10% tolerances) and note the results. By comparing the two cases, what conclusions can you draw?

EXERCISES

- Design a negative feedback amplifier that uses 100% feedback. What are its major characteristics?

QUESTIONS & PROBLEMS

1. To achieve negative feedback, the feedback signal must be:

    (a) in phase with the input signal.
    (b) 180° out of phase with the input signal.

2. What is a *direct-coupled* amplifier?

3. Why is the final stage of the amplifier of Figure 2.1 a Class B buffer?

4. Referring to the closed-loop amplifier of Figure 2.2, what *percentage* of the output voltage is fed back to the input?

5. How is the gain of the closed-loop amplifier of Figure 2.2 related to the values of the feedback resistors (RF1 and RF2)? How is it related to the percentage of feedback?

6. Why is the closed-loop amplifier of Figure 2.2 a better *buffer* than the open loop amplifier of Figure 2.1?

CHAPTER 3
Subcircuits
Creating Symbols

OBJECTIVES

- To create a subcircuit.
- To design and test a schematic that uses a subcircuit.

DISCUSSION

The three-stage amplifier of Figure 3.1 (reproduced from Chapter 2) might be called a "standard" circuit. It is a combination of fundamental components that performs a basic function (differential amplification and buffering).

In a large-scale schematic, such a circuit combination may appear many times. Rather than clutter the main schematic with the circuit details, it would be most convenient if we could represent the circuit as a single functional block (a *subcircuit*) and store it within a symbol library. Then, whenever we needed to incorporate such a device within our design, we could fetch it from the library as easily as we do with a diode or transistor. That is the purpose of this chapter.

To create a ready-to-use subcircuit, we follow a three-step sequence:

Step 1: Create the subcircuit MODEL and make it available *locally* (valid only within the schematic presently being edited).

Step 2: Create the subcircuit PART by designing the PART symbol, assigning the symbol a PART name, and storing it in a symbol library

Step 3: Associate the MODEL with the PART.

18 Part I Feedback

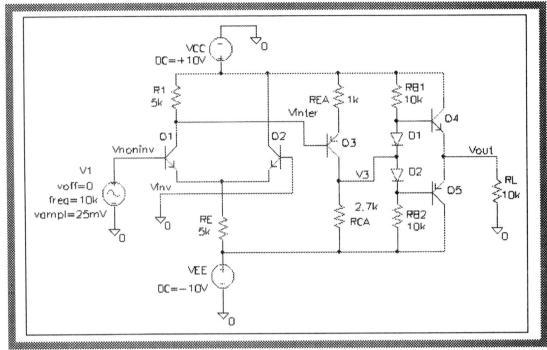

FIGURE 3.1

Three-stage amplifier
(open loop)

SIMULATION PRACTICE

Step 1: Create the MODEL

1. Bring back the three-stage differential amplifier/buffer of Figure 3.1 (from Chapter 2).

2. As shown in Figure 3.2, replace the input/output lines with the *interface port symbols* IF_IN and IF_OUT (from *port.slb*). To label each port symbol as shown, **DCLICKL** on each symbol and fill in.

3. Save (**File**, **Save As**) the schematic to file *opamp.sch*.

4. Perform **Tools**, **Create Subcircuit** to generate the subcircuit MODEL definition of Table 3.1 and store in file *opamp.sub* (in the same directory as *opamp.sch*). (Note that for subcircuits, the MODEL definition is a standard netlist of all the components and nodal connections making up the subcircuit.)

PSpice for Windows

Chapter 3 Subcircuits

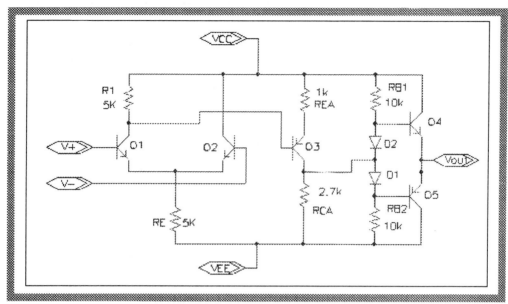

FIGURE 3.2
Substituting interface ports for I/O lines

Schematics Subcircuit .SUBCKT opamp V+ V- Vout VCC VEE	
Q_Q1	$N_001 V+ $N_0002 Q2N3904
R_R1	$N_0001 VCC 5K
R_RE	VEE $N_0002 5K
Q_Q4	VCC $N_0003 Vout Q2N3904
Q_Q5	VEE $N_0004 Vout Q2N3906
R_RCA	VEE $N_0005 2.715K
R_REA	VCC $N_0006 1K
Q_Q3	$N_0005 $N0001 $N---6 Q2N3906
Q_Q2	VCC V- $N_0002 Q2N3904
D_D1	$N_0005 $N_0004 D1N4148
D_D2	$N_0003 $N_0005 D1N4148
R_RB1	$N_0003 VCC 10K
R_RB2	VEE $N_0004 10K
.ENDS	opamp

TABLE 3.1
Subcircuit MODEL definition

PSpice for Windows

5. To make the MODEL definition (*opamp.sub*) available *locally* (valid only within the same directory as *opamp.sch*): **Analysis, Library and Include Files**, enter *opamp.sub* in the File Name field, **Add Library, OK**.

> To make the MODEL definition *global* (available to all schematics), place *opamp.sub* in directory *c:\msimev6x\lib*, where "x" is replaced with the PSpice version you are using (1, 2, 3, etc.) and perform: **Analysis, Library and Include Files**, enter *opamp.sub* in the File Name field, **Add Library*, OK**.

Step 2: Create the PART
---

6. To create a PART, we first create a symbol, assign the symbol a PART name (such as OPAMP), and store the PART in a symbol library (such as *mylib.slb*). This is accomplished as follows:

    File, Symbolize, type PART name OPAMP in *Save As* dialog box, **OK**, and save it to *mylib.slb* in directory *c:\msimev6x\lib*, where again x = the version you are using.

    > Hint: To save the symbol, the *Choose Library for Schematic Symbol* dialog box should have the entries shown below.
    >
    > | File Name | Directories |
    > |---|---|
    > | mylib.slb | c:\msimev6x\lib |

7. The next step is to edit the new PART symbol by bringing up the Symbol Editor: **File, Edit Library** (if necessary, **Yes** to save changes), **File, Open, DCLICKL** on "mylib.slb", **Part, Get, DCLICKL** on OPAMP to bring up the initial symbol.

    Move the reference designator and pin names to desired locations (as shown by the final schematic symbol of Figure 3.3).

8. To save the edited PART symbol back to the symbol library (*mylib.slb*): **File, Save, Yes** save to Part, **Yes** (if necessary) add to list of configured libraries. (Note: All symbol libraries [such as *mylib.slb*] are automatically global.)

PSpice for Windows

Chapter 3 Subcircuits

> **Step 3:** Associate the MODEL with the PART

9. To associate the MODEL definition (*opamp.sub*) with the PART symbol (in *mylib.slb*): **Part, Attributes, CLICKL** on PART, enter OPAMP in Value field, **Save Attr, CLICKL** on MODEL, enter OPAMP in Value field, **Save Attr, OK**. Note that the OPAMP part name appears on the symbol.

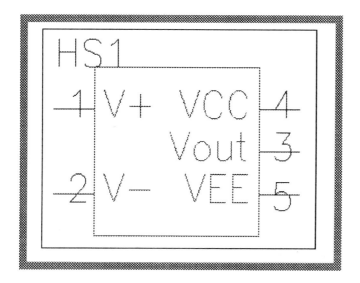

FIGURE 3.3
The OPAMP PART symbol

10. **File, Return to Schematics, Yes** save changes to Part, **Yes** save changes to library (to exit the symbol editor and return to Schematics).

Amplifier design, using the OPAMP subcircuit

11. Create a new schematic screen (**File, New**).

12. Fetch subcircuit PART OPAMP from "mylib.slb".

13. Add feedback components to create the circuit of Figure 3.4.

PSpice for Windows

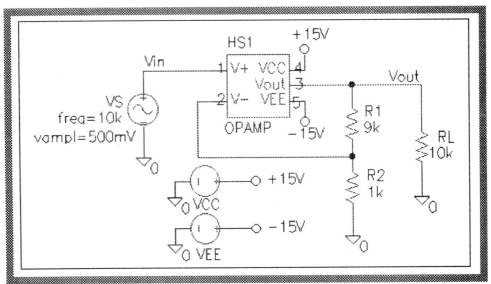

FIGURE 3.4
VCVS using
subcircuit opamp

14. Store the circuit in a file of your choosing, run PSpice (transient mode), and test the circuit. (Be sure to set VCC and VEE to +15V and -15V.)

    Is the voltage gain close to the expected value of 10? (See chapter 2, step 4.)

    Yes   No

15. To show that a subcircuit block is really a hierarchical module (see chapter 29, volume I), select block OPAMP, **DCLICKL** (or, **Navigate, Push**), and note that the block contents are displayed.

    To view an arbitrary waveform (such as the collector of Q3), drop a voltage marker at this location (or find and select alias variable V(HS1.Q3:c)).

Advanced activities

16. To demonstrate the ability to plot internal variables, use the hierarchy "dot" technique to plot the voltage at the collector of Q3. [Hint: Find alias variable V(HS1.Q3:c)]

17. Cascade two op amp circuits (of the type shown in Figure 3.4) to generate an overall gain of approximately 100.

PSpice for Windows

Chapter 3 Subcircuits

EXERCISES

- Create and test a power supply subcircuit based on the design of Chapter 9 (Volume I).

- Based on the design of Figure 3.5, create and test a subcircuit called INVERTER. Determine its input/output characteristics and its propagation delay.

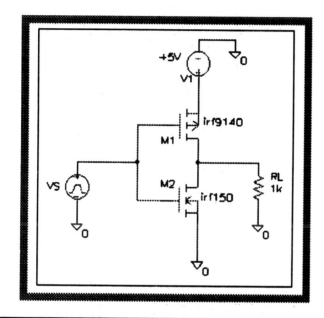

FIGURE 3.5
Basic inverter circuit

QUESTIONS & PROBLEMS

1. List several advantages of using subcircuits.

2. Can a circuit be constructed entirely of subcircuits?

3. Can we "push" into a subcircuit?

4. Referring to the trace variable *V(HS1.Q3:c)*, what does the dot mean?

5. What is the difference between a *local* and a *global* subcircuit?

CHAPTER 4
The Operational Amplifier
Specifications

OBJECTIVES

- To compare test results of the 741 op amp with its major specifications.

DISCUSSION

The "homemade" (subcircuit) op amp of the last chapter has a number of drawbacks. For starters, the gain and input impedance are too low, and the output impedance is too high. There are many other secondary characteristics we wish to improve.

In this chapter, we will examine a commercially available operational amplifier and see if it offers the improved characteristics we desire. Our choice is the 741 "workhorse" op amp, supplied within PSpice library *eval.slb*.

Specifications

The characteristics of any circuit, such as the 741 op amp, are known as *specifications*. The important specifications of the 741 op amp are listed in Table 4.1. Using PSpice we will verify each of these one at a time.

Of all the specifications listed, perhaps the most surprising is the very low value of the break frequency (10Hz). Even more surprising, it is done deliberately—in order to force the overall closed-loop op amp gain below one before phase shifts cause oscillations by way of positive feedback.

PSpice for Windows

741 specifications

• Input offset voltage	1mV
• Max/min output voltages	14V (15V supplies)
• Open loop voltage gain	200,000 (106dB)
• Input bias current	80nA
• Input offset current	20nA
• Short circuit output current	25mA
• Input impedance	2MEG
• Output impedance	75ohms
• CMRR	90DB (30,000)
• Slew rate	.5V/μs
• Break frequency	10Hz
• Frequency rolloff	20dB/Dec
• Gain-BW product	1MEGHz

TABLE 4.1

The 741 op amp specifications

SIMULATION PRACTICE

1. Draw the test circuit of Figure 4.1 and set the attributes as shown. (*ni* or + equals *non-inverting* and *i* or − equals *inverting*.) Note: pins 1 and 5 (OS1 and OS2), normally used for offset null adjustment, are inoperative and will be left floating.

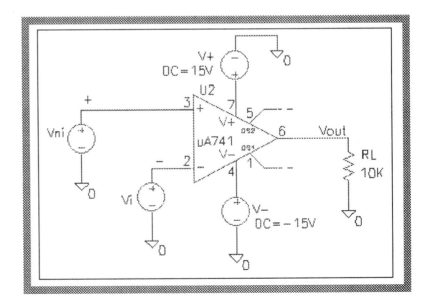

FIGURE 4.1

Op amp test circuit

Input offset voltage - *The voltage at either input that drives V_{OUT} to zero.*

2. Set Vi to zero (DC = 0). Sweep Vni between -200μV and +200μV, in steps of 1μV, and display the V_{OUT} curve of Figure 4.2. Determine the value of Vni that drives V_{OUT} to *approximately* zero. This value is V_{OFF}. Compare your result with the spec sheet value shown in Table 4.1.

V_{OFF} (PSpice) = _____ V_{OFF} (spec) = ___1 mV___

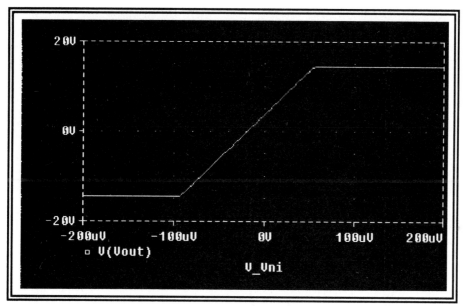

FIGURE 4.2
DC Sweep curves

Maximum/minimum V_{OUT} values - *Also known as the "rail" voltages, the maximum and minimum output voltages allowed for a particular power supply level (15V in our case).*

3. Using the graphical results from step 2, enter the "rail" voltages.

$V_{RAIL}+$ (PSpice) = _____ $V_{RAIL}+$(spec) = ___+14V___

$V_{RAIL}-$ (PSpice) = _____ $V_{RAIL}-$(spec) = ___-14V___

Open loop voltage gain - *The differential voltage gain without feedback.*

4. Again using the graphical results from step 2, enter the open-loop voltage gain. (Hint: Gain = $\Delta V_{OUT}/\Delta V_{IN}$, which is the slope of the V_{OUT}/V_{IN} curve.)

    A_{OL} (PSpice) = _____ A_{OL} (spec) = __200,000 (106dB)__

Input bias current - *The average value of the input currents (Vni and Vi grounded).*

5. Set Vni to DC zero (Vi remains at DC = 0). Using *bias point* techniques (examine the output file), determine Ini and Ii (the current through V_Vi and V_Vni), and use the equation below to calculate the input bias current.

$$I_{BIAS} \text{ (PSpice)} = \frac{Ini + Ii}{2} = \underline{\qquad} \qquad I_{BIAS} \text{ (spec)} = \underline{\text{80nA}}$$

Input offset current - *The difference between the input bias currents.*

6. Using the data from step 5, determine the input offset current.

    I_{OFFSET} (PSpice) = Ini - Ii = _____ I_{OFFSET} (spec) = __20nA__

Short circuit output current - *The output current when V_{OUT} is shorted.*

7. Set Vni to DC = +200µV (Vi remains grounded). Short the output to ground (set RL to a very small value) and use bias point techniques to determine the output current. (Hint: X_U1.vlim gives the output current.) When done, return RL to 10k.

    I_{SHORT} (PSpice) = _____ I_{SHORT} (spec) = __25mA__

Input impedance - *The total resistance between the inputs.*

8. Set Vni to DC = +1V (large enough to swamp out offset voltage and bias current levels) and leave Vi grounded. Using bias point techniques, measure Ini and determine the following:

    Z_{IN} (PSpice) = _____ Z_{IN} (spec) = __2MEG__

Output impedance - *The resistance viewed from the output terminal.*

9. Set Vni to ground and leave Vi grounded. Replace RL with VSRC and set to DC = +1V. Measure output current (X_U1.Vlim) and determine Z_{OUT}. When done, return VSRC to RL (10k).

    Z_{OUT} (PSpice) = _____ Z_{OUT} (spec) = __75Ω__

Common-mode rejection ratio (CMRR) - *Open-loop differential gain (A_{DM}) divided by common mode gain (A_{CM}).*

10. Short both inputs to Vni, as shown in Figure 4.3. Sweep Vni between +5V and -5V, display V_{out}, and determine A_{CM}. (Hint: A_{CM} equals the slope of the output curve.)

    A_{CM} = _____

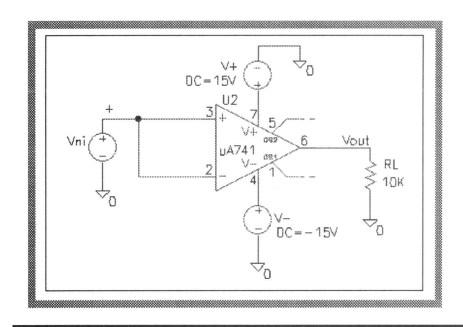

FIGURE 4.3
Common-mode configuration

11. Using the data gathered in steps 4 and 10, determine CMRR.

    CMRR (PSpice) = _____   CMRR (spec) = __90dB (30,000)__

Slew rate - *The slope of V_{OUT} to a step input signal.*

12. Ground the inverting lead (Vi). Redefine Vni for a step input of 0 to .1V and run a transient analysis from 0 to 50μs. (Use VPULSE and set *V1* = 0, *V2* = .1V, *td* = 10μs, and *tr* = 1ns. Note that for a simple step, *tf*, *pw*, and *per* need not be specified.)

13. Display V_{OUT} (to 50μsec) and determine its slew rate (slope) in volts/μsec.

    Slew rate (PSpice) = _____   **Slew rate (spec)** = __.5V/μs__

Frequency response - *The gain versus the frequency (Bode plot).*

14. Set Vni to AC = 50μV (use of VPULSE okay) and leave Vi grounded. Perform a logarithmic AC sweep from 1Hz to 10MEGHz, and generate the graph of Figure 4.4. Answer the following:

    (a) What is the break frequency?

    F_B **(PSpice)** = _____ F_B **(spec)** = __10Hz__

    (b) What is the rolloff? (<u>Hint</u>: Measure ΔVdB between 1KHz and 10KHz.)

    Rolloff (PSpice) = _____   **Rolloff (spec)** = __20dB/Dec__

    (c) What is the *gain-bandwidth product*? (<u>Hint</u>: measure the *unity gain frequency*, the frequency at which the gain is 1, or 0dB).

    Gain-BW (PSpice) = _____   **Gain-BW (spec)** = __1MEGHz__

    (d) Pick several points along the curve and determine A x f (gain times frequency). What generalization can you make about this value (known as the *gain/bandwidth product)*? [Note: A(regular) = $10^{A(dB)/20}$]

Advanced activities

15. Using the internal design of subcircuit OPAMP (Figure 3.1 from Chapter 3), find experimentally (PSpice) the value of capacitance placed between the base and collector of transistor Q3 that would cause the circuit to break at 10Hz. What does this tell us about how the designers of the 741 caused it to break at 10Hz?

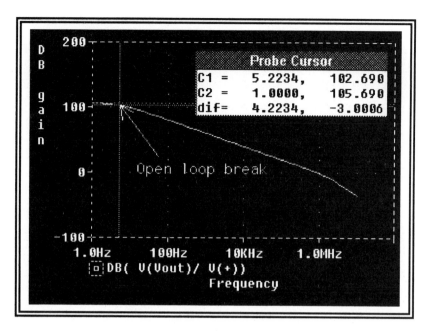

FIGURE 4.4
Open-loop Bode plot

EXERCISES

- Determine the specifications of the LM324 and compare with the UA741.

- Design a *comparator* circuit in which V_{OUT} = +15V for all V_{IN} greater than +5V and V_{OUT} = -15V for all V_{IN} +5V or less.

QUESTIONS & PROBLEMS

1. Is the circuit of Figure 4.1 in the *open-loop* or *closed-loop* configuration?

2. For many applications, why is it possible to ignore many of the specifications of this experiment (such as *input offset current*)?

3. How do the specifications for A, Z_{IN}, and Z_{OUT} for the 741 differ from the characteristics of the amplifier of the previous experiment?

4. Why does the 741 break at such a low frequency? (Hint: What happens in a feedback circuit if high-frequency phase shifts change negative feedback to positive feedback when the overall gain is also positive?)

CHAPTER 5
The Non-Inverting Configuration
VCVS & VCIS

OBJECTIVES

- To determine the characteristics of the VCVS and VCIS op amp configurations.

DISCUSSION

A common form of negative feedback is the VCVS (Voltage-controlled-voltage-source) mode of Figure 5.1.

> *A voltage-controlled-voltage-source is a device in which the output voltage is directly proportional to the input voltage, regardless of the load.*

It is useful to think of the VCVS configuration as a *voltage-to-voltage transducer*. That is, there is a linear one-to-one correspondence between the input voltage and the output voltage—regardless of load value. In other words, give me an input voltage, and I will give you the corresponding output voltage—the load (or output current) doesn't matter.

The transfer function

The *transfer function* is the ratio of output over input. For the VCVS configuration of Figure 5.1, both the input and output are voltage, and therefore the transfer function is its *voltage gain (A)*. Of special importance to the designer, the VCVS transfer function depends primarily on the feedback resistors (RF1 and RF2).

PSpice for Windows

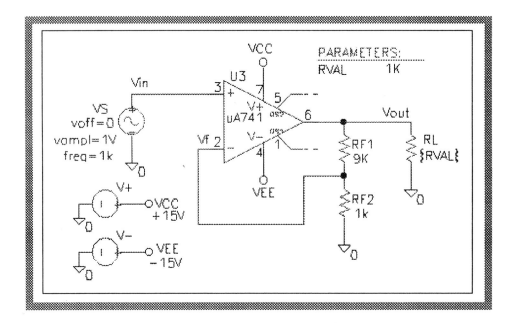

FIGURE 5.1

VCVS configuration

To calculate the closed-loop transfer function (A_{CL}) for the VCVS amplifier, we refer to the Thevenized version of Figure 5.2. We begin by writing the following equations:

$$V_{OUT} = A_{OL} \times (V_{IN} - V_F)$$

$$V_F = V_{OUT} \times (RF2/(RF1 + RF2))$$

Solving these equations simultaneously yields:

$$A_{CL} = \frac{1}{1/A_{OL} + \beta}$$

where β = feedback ratio = RF2/(RF1 + RF2)

The real versus ideal case

When performing theoretical calculations in the closed-loop (CL) case, a great simplification results if we assume the *ideal* (rather than *real*) case for the open-loop (OL) op amp specifications. *We will find that the errors that result from basing our calculations on the ideal case are very tiny.*

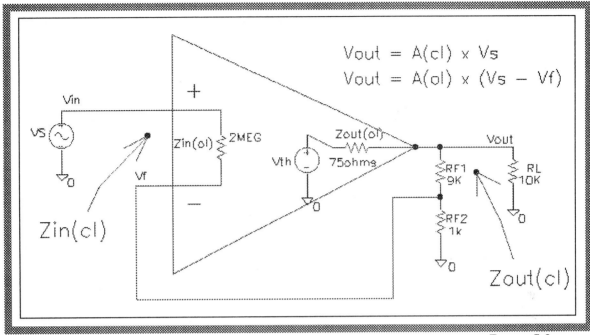

Figure 5.2
Thevenized VCVS circuit

Listed below are the "real" and "ideal" open-loop "big three" parameters for the 741 op amp.

Open loop parameters

Item	Real	Ideal
A_{OL}	200,000	∞
Zin_{OL}	2MEG	∞
$Zout_{OL}$	75ohms	0

The ideal transfer function

As an example of the use of an ideal parameter, we substitute the ideal value for open-loop voltage gain (∞) into the equation developed earlier and generate the *ideal* closed-loop voltage gain:

$$A_{CL} = \frac{1}{1/A_{OL} + \beta} = \frac{1}{1/\infty + \beta} = \frac{1}{\beta} = 1 + \frac{RF1}{RF2}$$

PSpice for Windows

A perfect amplifier

When all the characteristics of the VCVS configuration are put together, what we have is a nearly perfect voltage amplifier. It gives us ease of design, DC or AC operation, high Z_{IN}, low Z_{OUT}, high bandwidth, and low distortion. It is called *non-inverting* because V_{IN} and V_{OUT} are in phase.

The shadow rule

During normal operation, negative feedback assures us that the inverting input will track (follow) the non-inverting input. This very useful analysis and troubleshooting aid will be known as the "shadow rule" (named after the famous Vaudeville act "Me and My Shadow").

The shadow rule is a direct consequence of negative feedback. *In the ideal case*, the op amp will always generate an output voltage that will drive the input differential (error) voltage to zero. *Therefore, the voltage at the inverting input must track (shadow) the voltage at the non-inverting input.*

The VCIS configuration

The other major form of the non-inverting configuration is the VCIS (*voltage-controlled-current-source*) of Figure 5.3.

> *A voltage-controlled-current-source is a device in which the output current is directly proportional to the input voltage, regardless of the load.*

As before, think of the VCIS configuration as a *voltage-to-current transducer*. There is a linear one-to-one correspondence between the input voltage and the output current—regardless of load value. Give me an input voltage, and I will give you the corresponding output current—the load (or output voltage) doesn't matter.

For this VCIS configuration, the transfer function is I_{OUT}/V_{IN}, which gives units of *transconductance*. Making use of the "shadow rule," and assuming the ideal case, the closed-loop transfer function is calculated as follows:

$V_{IN} = V_F$ $I_{OUT} = V_F/RF$

transconductance = $I_{OUT}/V_{IN} = 1/RF$

PSpice for Windows

Chapter 5 The Non-Inverting Configuration

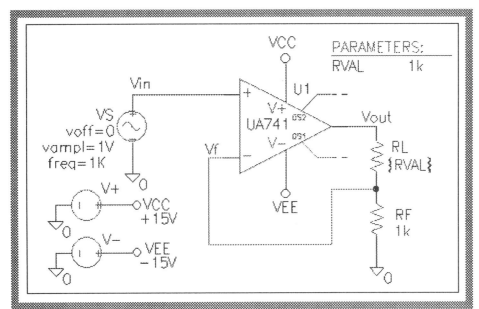

FIGURE 5.3
The VCIS non-inverting configuration

Because the load resistor (RL) is not referenced to ground, the VCIS configuration is seldom used. (It will appear only as an advanced activity.)

SIMULATION PRACTICE

VCVS

1. Draw the VCVS circuit of Figure 5.1, and set up the load for parametric analysis. (To draw the power supply connections, see *Schematics Note 5.1*.)

2. Using equations and open-loop parameters developed in the discussion, calculate both the *real* and *ideal* voltage gain.

    A_{REAL} = _____ A_{IDEAL} = _____

3. Based on the results of step 2, are we justified in using the ideal case to calculate the voltage gain?

    Yes   No

PSpice for Windows

> ## Schematics Note 5.1
> *How do I simplify power connections?*
>
> To simplify power connections (as shown in Figure 5.1), first move the DC power sources (VSRC) off to the side of the schematic. To place the *interconnect bubbles*: **Draw, Get New Part, Browse, port.slb, BUBBLE, OK, DRAG** to desired location, **CLICKL** (repeat to place multiple bubbles), **CLICKR** to abort. Position bubbles as desired. **DCLICKL** on each bubble to bring up the *Set Attribute Value* dialog box, enter desired bubble name (making sure the corresponding bubbles match), **OK**.

4. To prove that the circuit of Figure 5.1 is a *voltage-controlled-current-source* (VCVS), consider the following:

   > *For a sine wave input voltage, a circuit is a VCVS if the output voltage is a perfect sine wave whose amplitude is independent of the load (RL).*

5. Generate the graph of Figure 5.4, which shows V_{IN} and V_{OUT} for various values of RL from 1kΩ to 5kΩ.

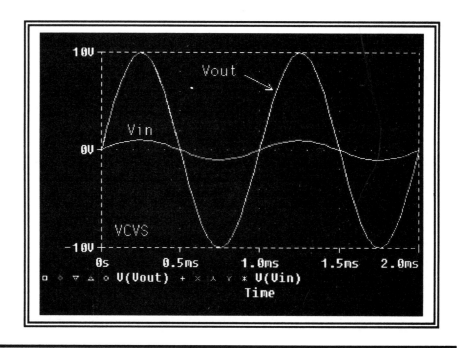

FIGURE 5.4

VCVS test results

Chapter 5 The Non-Inverting Configuration

6. Using the waveforms of Figure 5.4, determine the transfer function (voltage gain) as accurately as you can (use the cursor).

    A(PSpice) = _____

    Is this "experimental" value close to the calculated values of step 2?

    Yes   No

7. Does the graph of Figure 5.4 seem to prove that the circuit is indeed a near-perfect VCVS? (Does it appear that V_{OUT} is a sine wave whose amplitude is ten times V_{IN} and independent of RL?)

    Yes   No

8. Zoom way in on any portion of the V_{OUT} curve and note that there are actually five closely spaced curves. Is it now fair to say that the VCVS circuit is a *very good* (but not *perfect*) VCVS with a voltage gain of 10?

    Yes   No

Input impedance

9. Determine the input impedance [$Z_{IN}(CL)$] for each case below. [Note: the *real* equation shown below for $Z_{IN}(CL)$ came from an analysis of Figure 5.2.]

    (a) The calculated *real* case [assume $Z_{IN}(OL)$ = 2MEG].

    $Z_{IN}(CL)$ (real) = $(1 + \beta A_{OL}) \times Z_{IN}(OL)$ = _____

    (b) The calculated *ideal* case. (Hint: substitute A_{OL} = infinity in the above equation)

    $Z_{IN}(CL)$ (ideal) = _____

    (c) Using PSpice. (Add I_{IN} to Figure 5.4, and be sure to use the peak value in your calculations.)

    $Z_{IN}(CL)$ (PSpice) = _____

10. Comparing the three values of step 9, are we justified in using the ideal case for determining the closed-loop input impedance?

    Yes   No

PSpice for Windows

Output impedance

11. Determine the closed-loop output impedance [$Z_{OUT}(CL)$] for each case below.

    (a) The calculated *real* case. [Assume $Z_{OUT}(OL)$ = 75 ohms.]

    $$Z_{OUT}(CL)\ (real) = \frac{Z_{OUT}(OL)}{1 + A_{OL}\beta} = \underline{\hspace{2cm}}$$

    (b) The calculated *ideal* case. [Hint: looking at the real case equation, what is $Z_{OUT}(CL)$ when A_{OL} = infinity?].

    $Z_{OUT}(CL)\ (ideal)$ = _____

    (c) Using PSpice. (Measure V_{OUT} with and without a load and use algebra. Or, substitute VSRC for RL, ground V_{IN}, and measure I_{OUT}—making sure that $I_{OUT} < I_{MAX}$.)

    $Z_{OUT}(CL)\ (PSpice)$ = _____

12. Comparing the three values of step 11, are we justified in using the ideal case for determining the closed-loop output impedance?

    Yes   No

Harmonic distortion

13. Determine the total percent harmonic distortion of the output signal.

    Total HD_{CL} (PSpice) = _____

    Compared to the open-loop case of Chapter 2, is the value small?

    Yes   No

The shadow rule

14. Plot graphs of V_{NI} (V_{IN}) and V_I (V_F). Does the inverting input *approximately* track the non-inverting input? (If you wish, zoom in on the curves.)

    Yes   No

Chapter 5 The Non-Inverting Configuration

Bode plot

15. Assign attribute *AC = 1V* to VS and set up the AC Sweep mode.

16. Determine the closed-loop bandwidth (BW_{CL}) for each case below.

    (a) The calculated *real* case.

    The bandwidth equals the frequency at which A_{CL} breaks (drops 3dB). As calculated below, this happens when A_{OL} drops by 27.65dB.

    $$\text{when } A_{CL} = 7.07 \, (70.7\% \text{ of } 10) = \frac{A_{OL}}{1 + A_{OL}(1/10)}$$

    $A_{OL} = 24.129 \, (27.65\text{dB})$

    Referring to the open-loop Bode plot of Figure 5.5 (reproduced from Chapter 4), obtain the approximate frequency when A_{OL} is 27.65dB and fill in below.

    BW_{CL} (real) = _____

    (b) The calculated *ideal* case.

    We skip this case because there is no ideal value. (The open-loop bandwidth is deliberately limited to approximately 10Hz in order to avoid high-frequency phase shifts from turning the circuit into a positive feedback oscillator.)

    (c) Using PSpice.

    BW_{CL} (PSpice) = _____

17. Does the calculated value of the closed-loop bandwidth (16a) *approximately* equal the experimental value (16c)?

    Yes   **No**

18. For the VCVS configuration, generate the family of Bode plot curves (AC sweep) of Figure 5.6 by sweeping RF1 over a range of values (such as 1k, 9k, and 99k).

    (a) For all curves, does the ideal formula for midband gain (1 + RF1/RF2) approximately hold?

    Yes   **No**

    (b) For all curves, is the gain/bandwidth product *approximately* a constant? If so, what is this value?

    $A_{CL} \times BW_{CL}$ = _____

PSpice for Windows

42 Part I Feedback

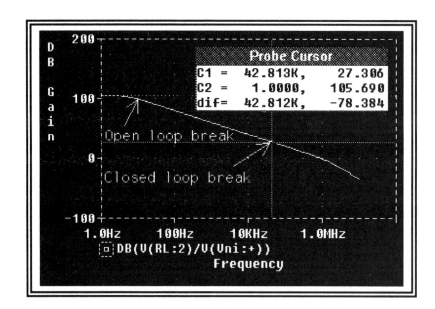

FIGURE 5.5

Open loop Bode plot

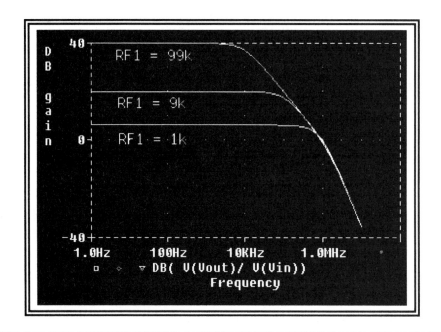

FIGURE 5.6

Family of Bode plot curves

PSpice for Windows

Advanced activities

19. Using transient analysis on the VCVS configuration, compare closed-loop parameters A, Z_{IN}, Z_{OUT}, and HD at 1kHz (before the break frequency) with the same values at 10MHz (well after the break frequency).

    (a) What generalizations can you make?

    (b) Does the "magic" of negative feedback disappear at high frequencies?

20. Perform a complete analysis of the VCIS (*voltage-controlled-current-source*) of Figure 5.3. Determine any or all of the following and compare to the VCVS of Figure 5.1:

    (a) Graph of I_{OUT}(CL) versus $V_{in(CL)}$ for various values of RL from 1kΩ to 10kΩ. (Is the circuit a VCIS?)

    (b) The transfer function (transconductance).

    (c) Z_{IN}(CL) and Z_{OUT}(CL).

    (d) Harmonic distortion (HD_{CL}).

    (e) Bandwidth (CL)

21. Compare the results of step 20 with the ideal characteristics of a VCIS. Would we be justified in using the ideal case for most applications?

EXERCISES

- Investigate the output properties of the *buffered* VCVS configuration of Figure 5.7. [How are the closed loop parameters (A, Z_{IN}, Z_{OUT}, BW, or HD) affected? How low can RL go before current limiting is reached?]

- Investigate the properties of the "perfect" diode of Figure 5.8.

- Investigate the properties of the "perfect" clamper of Figure 5.9. (Change Vref and note the results.)

44 Part I Feedback

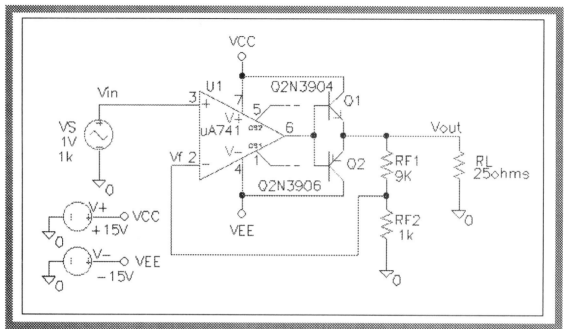

FIGURE 5.7

Buffered VCVS

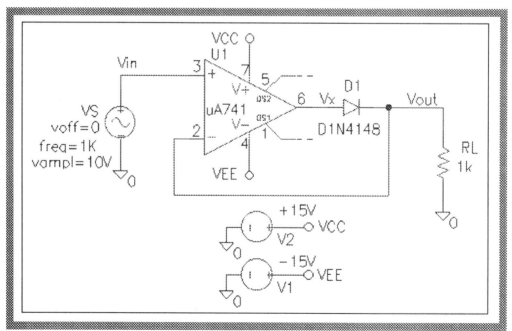

FIGURE 5.8

Ideal diode

PSpice for Windows

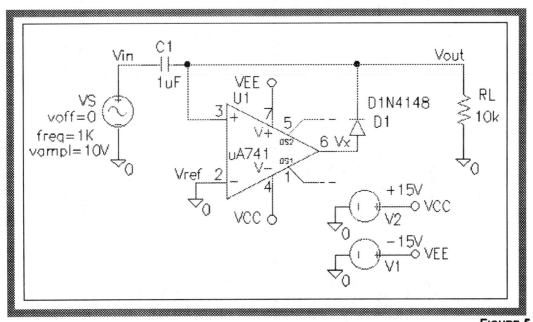

FIGURE 5.9
Ideal clamper

QUESTIONS & PROBLEMS

1. Based on all the data and results of this experiment, why can the configuration of Figure 5.1 be called a VCVS (voltage-controlled-voltage-source)?

2. What is the difference between *open loop* and *closed loop*?

3. For the VCVS circuit, why is there no low-frequency breakpoint?

PSpice for Windows

4. Referring to the circuit of Figure 5.1, approximately what percent error in gain results from assuming the ideal case?

5. Why is the VCVS configuration called *non-inverting*?

6. Based on the results of this chapter:

    (a) What is the ideal Z_{IN} for a voltage-controlled (VC) input?

    (b) What is the ideal Z_{OUT} for a voltage source (VS)?

    (c) What is the ideal Z_{OUT} for a current source (IS)?

CHAPTER 6
The Inverting Configuration
ICVS & ICIS

OBJECTIVES

- To determine the characteristics of the ICVS and ICIS op amp configurations.

DISCUSSION

The four basic closed-loop configurations are listed below. The first two were the topic of the previous chapter; the bottom two are the subject of this chapter.

- VCVS   (voltage-controlled-voltage-source)
- VCIS   (voltage-controlled-current-source)
- ICVS   (current-controlled-voltage-source)
- ICIS   (current-controlled-current-source)

Because of the lessons we learned with the first two configurations, we can more quickly cover the remaining two. For example, we already know the following:

- The ideal case can be used for most applications with very little error.
- Ideal transfer functions are easy to develop when using the shadow rule.
- A voltage-controlled (VC) input has an ideal Z_{IN} of infinity, and a current-controlled (IC) input has an ideal Z_{IN} of zero.
- A voltage-source (VS) has an ideal Z_{OUT} of 0, and a current-source (IS) has an ideal Z_{OUT} of infinity.

PSpice for Windows

The ICVS configuration

The most versatile negative feedback configuration for an op amp is the ICVS (current-controlled-voltage-source) of Figure 6.1.

> *A current-controlled-voltage-source is a device in which the output voltage is directly proportional to the input current, regardless of the load.*

Think of the ICVS configuration as a *current-to-voltage transducer*, with a linear one-to-one correspondence between the input current and the output voltage—regardless of load value. Give me an input current, and I will give you the corresponding output voltage—the load doesn't matter.

Virtual ground

The great versatility of the ICVS configuration is closely related to the creation of the *virtual ground* at the inverting input. (A *virtual ground* is at zero volts, but is not directly connected to ground.)

To see why a virtual ground is created, remember the *shadow rule*. With the non-inverting input at zero volts (grounded), an ideal op amp will always drive the input differential (error) voltage to zero—which means the inverting input must be driven to zero volts.

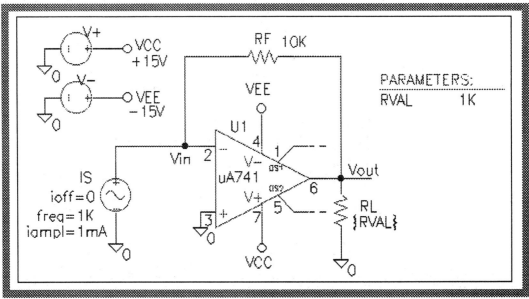

FIGURE 6.1
ICVS configuration using true current source (IS)

The ICVS transfer function

For the ICVS configuration of Figure 6.1, the input is current and the output is voltage, and therefore the transfer function is its *transresistance.* In the ideal case, the transfer function depends only on the single feedback resistor (RF). (This time, unlike the previous chapter, it will be the student's responsibility to determine the transresistance equation.)

A handy current source

Because the inverting input is at zero volts (a *virtual ground*), we can substitute a voltage-source/resistor combination for the true current source (as shown by Figure 6.2). This configuration is called *inverting* because Vout is 180° out of phase with VS.

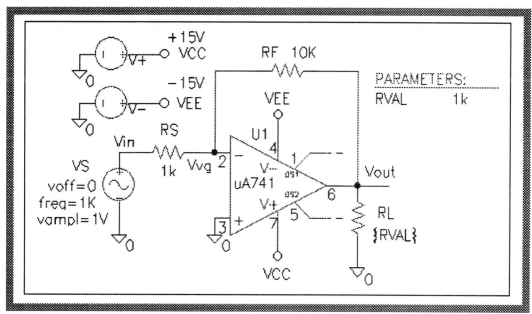

FIGURE 6.2
Using VS/RS as current source

If the input is taken as VS, the configuration of Figure 6.2 is often called a voltage amplifier (although it does not have the desirable high Z_{IN} characteristics of the VCVS configuration of the last chapter).

PSpice for Windows

The ICIS configuration

The other major form of the inverting configuration is the ICIS (*current-controlled-current-source*) of Figure 6.3.

> *A current-controlled-current-source is a device in which the output current is directly proportional to the input current, regardless of the load.*

As before, think of the ICIS configuration as a *current-to-current transducer*, with a linear one-to-one correspondence between the input current and the output current—*regardless of load value*. Give me an input current, and I will give you the corresponding output current—the load doesn't matter.

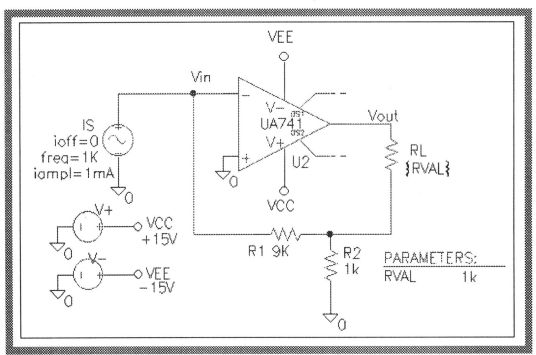

FIGURE 6.3

The ICIS inverting configuration (IS input)

The ICIS transfer function

For the ICIS configuration of Figure 6.3, both the input and output are current, and therefore the transfer function is *current gain* (β). In the ideal case, the transfer function depends only on feedback resistors (RF1 and RF2). As with the VCIS configuration of the last chapter, the ICIS configuration is seldom used because voltage usually is the preferred output variable.

SIMULATION PRACTICE

The ICVS configuration

1. Draw the ICVS circuit shown in Figure 6.1.

2. Using the ideal case, derive an ICVS *transresistance* equation for V_{OUT} versus I_{IN} in terms of RF. (*Hint*: Use the *shadow rule*.)

    V_{OUT} = (   ) × I_{IN}

3. Using the equation you developed in step 2, calculate the transresistance gain for the ICVS circuit of Figure 6.1.

    transresistance = V_{OUT}/I_{IN} = _____

4. To prove that the circuit of Figure 6.1 is an ICVS, consider the following:

    > For a sine wave input current, a circuit is an ICVS if the output voltage is a perfect sine wave whose amplitude is independent of the load (RL).

5. Generate the graph of Figure 6.4, which shows I_{IN} and V_{OUT} for various values of RL from 1kΩ to 5kΩ.

6. Based on the waveforms of Figure 6.4, determine the transfer function value (transresistance) as accurately as you can.

    transresistance (PSpice) = _____

    Is this "experimental" value close to the calculated value of step 3?

    Yes   No

52 Part I Feedback

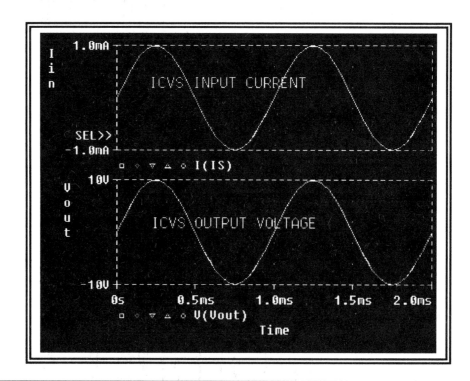

FIGURE 6.4

Plot of V_{out} versus I_{in}

7. Zoom way in on any portion of Vout and note the five closely spaced curves. Is it fair to say that the circuit of Figure 6.1 is a very good (but not perfect) ICVS with a *transresistance* of 10kΩ?

    Yes   **No**

8. Realizing that the input of the ICVS circuit of Figure 6.1 is a virtual ground and the output is a voltage source, *ideally* what would you expect the circuit's Z_{IN} and Z_{OUT} to be?

    Z_{IN} (ideal) = _____ Z_{OUT} (ideal) = _____

9. Measure Z_{IN} and Z_{OUT} using PSpice and compare to the results of step 8. (Hint: To determine Z_{OUT}, measure V_{OUT} with and without a load)

    Z_{IN} (PSpice) = _____ Z_{OUT} (PSpice) = _____

10. Comparing steps 8 and 9, are we justified in assuming ideal characteristics for the op amp?

    Yes   **No**

PSpice for Windows

11. Is the virtual ground at *approximately* zero volts (less than 20mV)?

    Yes **No**

The inverting voltage amplifier

12. Modify your circuit to generate the amplifier configuration of Figure 6.2, which uses a voltage-source/resistor combination for a current source.

13. Determine the circuit's voltage gain (V_{OUT}/V_S) using:
    (a) The calculated ideal case.

    A(Ideal) = RF / RS = _____

    (b) Using PSpice (transient mode).

    A(PSpice) = _____

    (c) Are the ideal and experimental (PSpice) values approximately the same?

    Yes **No**

    (d) Are the input and output voltages (V_{IN} and V_{OUT}) 180° out of phase?

    Yes **No**

Advanced activities

14. Determine the characteristics of the ICVS configuration of Figure 6.1 to the left and right of the breakpoint. Summarize your results.

15. Perform a complete analysis of the ICIS (*current-controlled current source*) of Figure 6.3. Determine any or all of the following and compare to the ICVS of Figure 6.1:

    (a) Graph of I_{OUT} versus I_{IN} for various values of RL from 50Ω to 500Ω. (Is the circuit a ICIS?)

    (b) The transfer function (transconductance).

    (c) Z_{IN} and Z_{OUT}.

    (d) Harmonic distortion (HD).

    (e) Bandwidth (BW).

PSpice for Windows

16. Compare the results of step 15 with the ideal characteristics of an ICIS. Would we be justified in using the ideal case for most applications?

17. Show how to use the ICVS configuration to convert the input current from the collector of a transistor (a current source) to output voltage.

18. Using the equation below, at what frequency is the maximum slope of a sine wave of 10V amplitude equal to the slew rate of the op amp (.5V/μsec)? [Hint: the maximum value of cos() = 1.]

    slope = $\Delta 10\sin(2\pi ft) / \Delta t = 20\pi f\cos(2\pi ft)$ = .5V/μsec

    f(max) = _____

    View the output waveform for a frequency approximately 20% above this frequency. Do you notice any distortion during the times of greatest slope (near the axis)?

EXERCISES

* Draw the *summing* amplifier of Figure 6.5 and show that the output voltage waveform is the algebraic sum of the two input voltage waveforms.

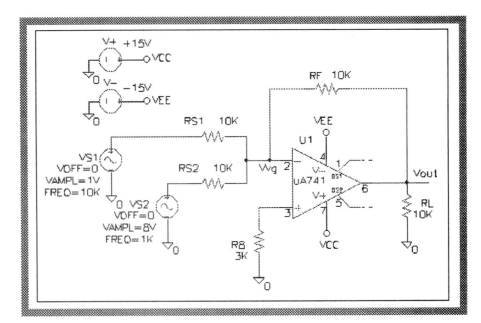

FIGURE 6.5

The summing amplifier

- Determine the transfer function of the *difference amplifier* of Figure 6.6 and test your equation with PSpice.

- For the *logarithmic amplifier* of Figure 6.7, what is the relationship between Vout and Vin? (Hint: is the plot of Vout versus LOG(V_{IN}) a straight line?)

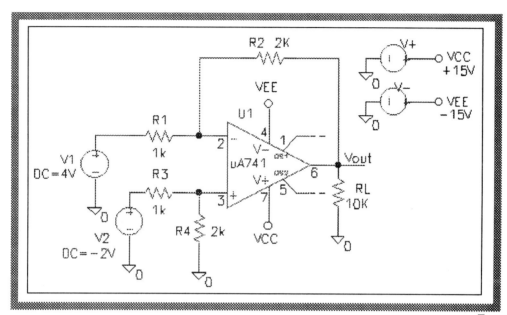

FIGURE 6.6
Difference amplifier

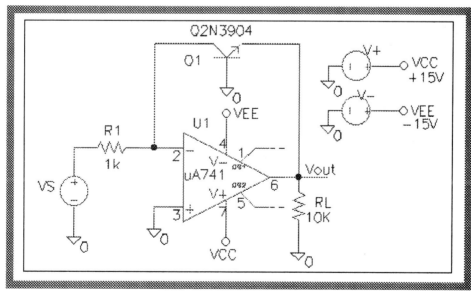

FIGURE 6.7
Logarithmic amplifier

PSpice for Windows

QUESTIONS & PROBLEMS

1. What is the difference between *ground* and *virtual ground*?

2. When used as a *voltage amplifier*, why is the VCVS configuration of Chapter 5 considered superior to the ICVS configuration (Figure 6.2) of this chapter?

3. A certain ICVS amplifier has a gain/bandwidth product of 10MEG. If the gain is 80, what is the bandwidth?

4. Regarding the summing amplifier of Figure 6.5:
   (a) Could the circuit be used by an analog computer to perform the *add* process?

   (b) How does the summing amplifier eliminate *crosstalk* (V_{s1}'s circuit affecting V_{s2}'s circuit and vice versa.) ?

PSpice for Windows

5. Place type VCVS, VCIS, ICVS, or ICIS after each of the following:

Z_{IN}	Z_{OUT}	Type
0	0	
0	infinity	
infinity	0	
infinity	infinity	

CHAPTER 7
Op Amp Integrator/Differentiator
Calculus

OBJECTIVES

- To review the fundamentals of calculus.
- To perform integration and differentiation using both RC and op amp circuits.

DISCUSSION

We are well aware of the relationships between distance (D), velocity (V), and time (t):

$$D = V \times t \qquad\qquad V = D / t$$

However, the well-known *multiply* and *divide* operators ($\times$ and $/$) only work if velocity is a *constant*. If velocity is a *variable* (a changing function of time), we must turn to the techniques of *calculus*, developed in the 16th century by Isaac Newton—we must *integrate* and *differentiate*.

$$D(t) = \int_0^t v(t)dt \qquad\qquad V(t) = \frac{dD(t)}{dt}$$

        Integration   **Differentiation**

As shown in Figure 7.1, integration is performed graphically by taking the accumulated area under the curve, and differentiation is performed graphically by taking the slope of the curve.

PSpice for Windows

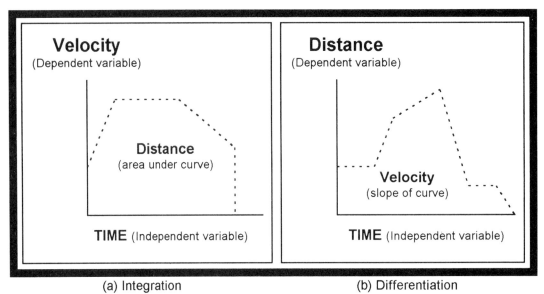

FIGURE 7.1
Graphical methods of
integration and differentiation
(a) Integration
(b) Differentiation

Electronic integration and differentiation

To see how integration and differentiation can be performed electronically, we note the relationship between current and charge:

$$Q(t) = \int_0^t I(t)dt \qquad I(t) = \frac{dQ(t)}{dt}$$

However, it is more convenient if both input and output can be measured in voltage. Therefore, we use a resistor as a current-to-voltage transducer (V = IR), and we take advantage of the relationship between capacitor voltage and charge (Q = CV) to generate:

$$Vout(t) = \frac{1}{RC}\int_0^t Vin(t)dt \qquad Vout(t) = RC\frac{dVin}{dt}$$

Based on these equations, we construct the simple RC-based circuits of Figure 7.2.

PSpice for Windows

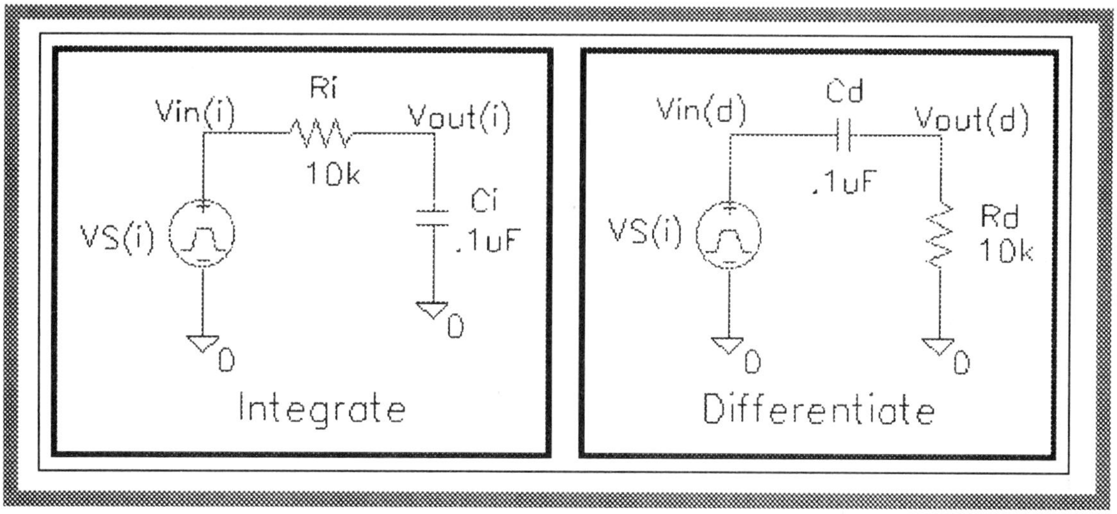

Chapter 7 Op Amp Integrator/Differentiator

FIGURE 7.2
RC integrator and differentiator

Unfortunately the simple RC-based circuits of Figure 7.2 do not perform perfect mathematical integration and differentiation because the voltage in the capacitor feeds back and affects the current in the resistor. Therefore, to carry out mathematically correct integration and differentiation, we must uncouple the resistor from the capacitor by maintaining the junction at zero volts (a *virtual ground*!).

Adding op amps to create the virtual grounds, we arrive at the circuits of Figures 7.3 and 7.4.

SIMULATION PRACTICE

Integrator

1. Draw the op amp integrator shown in Figure 7.3, and program VPULSE (VS) to generate the input waveform of Figure 7.5. The rise and fall times (td and tr) can be any small value, such as 10ns.

    To make the output predictable, note that we initialize capacitor C to zero.

PSpice for Windows

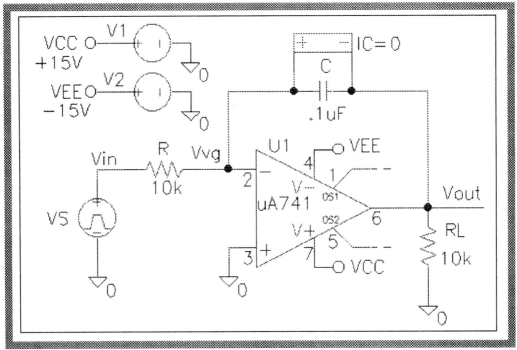

Figure 7.3

Op amp integrator

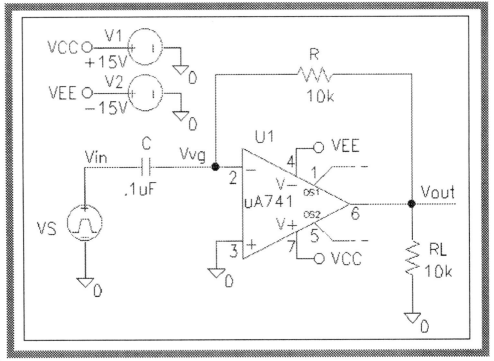

Figure 7.4

Op amp differentiator

PSpice for Windows

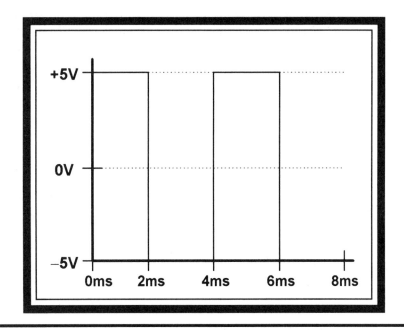

FIGURE 7.5
Integrator waveforms

2. Predict the output waveform and sketch your answer on the graph of Figure 7.5. (Hint: the output amplitude profile is the accumulated area under the curve of the input waveform.)

3. Generate the output waveform using PSpice and add your curve to Figure 7.5. Are the two curves similar?

    Yes   No

   Suggestion: Initialize C to +5V and repeat.

Differentiator

4. Draw the op amp differentiator shown in Figure 7.4, and program VPULSE (VS) to generate the input waveform of Figure 7.6.

5. Predict the output waveform and sketch your answer on the differentiator graph of Figure 7.6. (Hint: The output waveform is equal to the slope of the input.)

6. Generate the output waveform using PSpice. (Be prepared for an unexpected result.)

PSpice for Windows

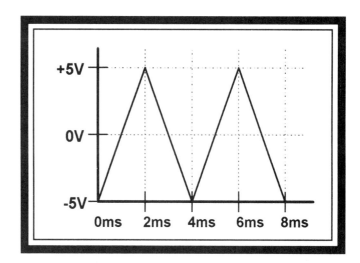

FIGURE 7.6
Differentiator waveforms

7. It is clear that significant noise has been produced. This is because a differentiator amplifies noise [the derivative (slope) of a high-frequency signal is greater than its amplitude].

   To solve the problem, add a small resistor to the input circuit (Figure 7.7) so the high-frequency noise will not "see" the capacitor.

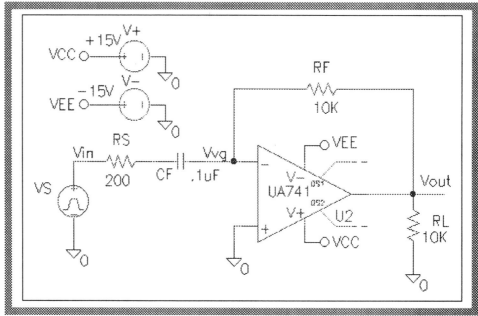

FIGURE 7.7
Differentiator with noise suppression

PSpice for Windows

8. Again generate the output waveform using PSpice and add your curve to Figure 7.6. Are the curves similar, and has the noise been greatly reduced?

    Yes   No

Advanced activities

9. For the input waveforms shown in Figure 7.8, predict the integrated and differentiated output waveforms and draw them on the graph.

10. For both the integrator of Figure 7.3 and the differentiator of Figure 7.7, change VS to VPWL and program the sources to match the waveforms of Figure 7.8. (See *Probe Note 7.1*.)

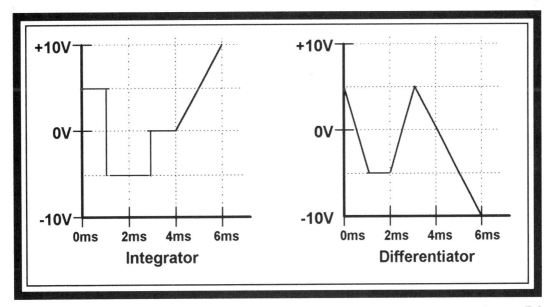

FIGURE 7.8
Input waveforms

11. Using PSpice, generate both the integrator and differentiator waveforms. Did your predictions match the actual case?

12. Using various waveforms, compare the output of the simple RC integrator and differentiator of Figure 7.2 with the op amp counterparts of Figures 7.3 and 7.7. How do they differ?

PSpice for Windows

> ## Probe Note 7.1
> ### *How do I create "custom" waveforms out of voltage segments?*
>
> Place a copy of VPWL (library *source.slb*) on your schematic. **DCLICKL** on the symbol to bring up the *Part Name* dialog box. Note the series of voltage and time pairs, where each pair represents a voltage/time node of the input waveform.
>
> For example, to generate the waveform of Figure 7.8, enter the following voltage/time pairs (be sure to **Save Attr** after each entry). :
>
> $t1 = 0ms \quad t2 = 1ms \quad t3 = 1.01ms \quad t4 = 3ms \quad t5 = 3.01ms \quad t6 = 4ms \quad t7 = 6ms$
> $v1 = +5V \quad v2 = +5V \quad v3 = -5V \quad v4 = -5V \quad v5 = 0V \quad v6 = 0V \quad v7 = +10V$
>
> For waveforms that have repetitive features, use source VPWL_ENH; for waveform values that are stored in a file, use VPWL_FILE.

13. The "compensated" integrator of Figure 7.9 uses resistor RF to prevent "DC buildup" (perhaps due to an offset voltage). Test the circuit over an extended range of time and over a range of pulse widths. Compare your results to the "uncompensated" integrator of Figure 7.3.

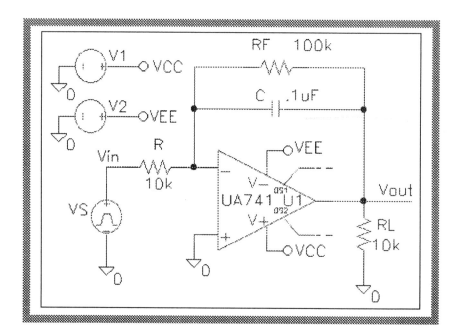

FIGURE 7.9

Compensated integrator

EXERCISES

* By cascading two differentiators, design a *double differentiator*. If the input is distance, what is the output? Also, test your design on various input waveforms.

* Making use of the circuit of Figure 7.10, test a "moon-landing simulator" that solves the equation below for an object of unity mass under control of thrust (T) and gravity (G).

    Using the transient mode, plot various "trajectories." (Change the gravity value, and experiment with various VSIN, VPULSE, and VPWL for Vthrust.)

$$D(t) = \iint (T(t) - G(t))dt^2$$

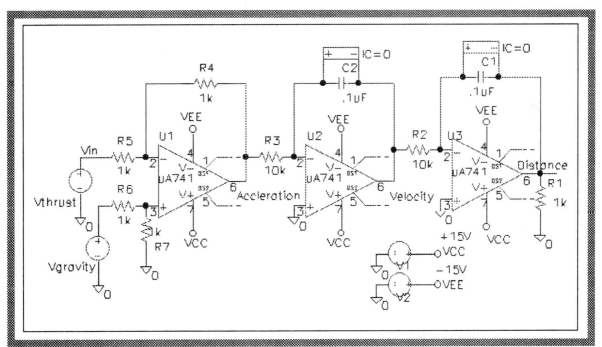

FIGURE 7.10
Moon-landing simulator

PSpice for Windows

QUESTIONS & PROBLEMS

1. What is the difference between *multiplication* and *integration*? Under what conditions does integration reduce to multiplication?

2. What is the difference between *division* and *differentiation*? Under what conditions does differentiation reduce to division?

3. When performing integration graphically, the output is the area under the curve of the input. Using this fact as a guide, how could you perform integration using an accurate scale and a pair of scissors?

4. Explain how to perform integration (over time) mechanically with the bucket and faucet of Figure 7.11. (Use the space on the next page.)

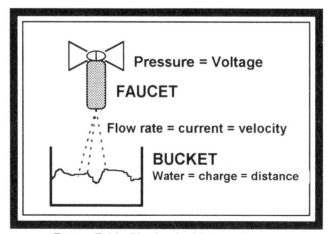

FIGURE 7.11 Mechanical integration

Chapter 7 Op Amp Integrator/Differentiator

5. Referring to the simple RC integrator of Figure 7.2, why is Vout equal to the integral of Vin only when the capacitor is nearly empty of charge?

6. Mathematically, under what input conditions would the output of an integrated waveform go up when the input was going down?

7. Mathematically, under what input conditions would the output of a differentiated waveform be negative when the input was positive?

8. Why is noise amplified by a differentiator, but not an integrator?

CHAPTER 8

Oscillators
Positive Feedback

OBJECTIVES

* To determine the conditions necessary for oscillation.
* To design and test sine and square wave oscillators.

DISCUSSION

Switching from negative to positive feedback requires only a tiny change in circuit configuration—but results in a huge change in operation.

For example, If our body temperature regulation mechanism changed from the stabilizing effects of negative feedback to the runaway effects of positive feedback, we would quickly succumb to a rapidly decreasing or increasing body temperature. When positive feedback is combined with negative feedback, the result is often oscillation.

Specifically, three conditions are necessary for oscillation:

1. **Closed-loop gain >= 1.**
2. **Positive feedback (loop phase shift = $360°$).**
3. **Initial spark (noise).**

The two major classifications of oscillators are *sine* and *square*. Within each category are numerous varieties and types. This chapter will highlight several popular oscillator configurations.

PSpice for Windows

SIMULATION PRACTICE

Sine wave oscillators

<u>Phase-shift</u>

1. Draw the *phase-shift* oscillator of Figure 8.1, a good choice for low-frequency applications.

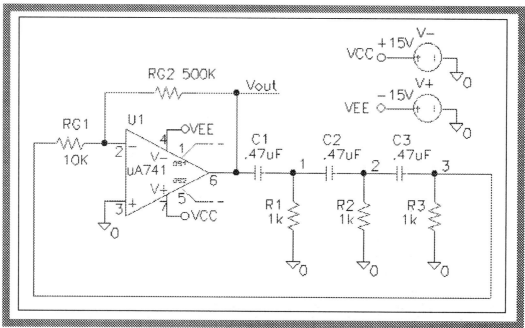

FIGURE 8.1
The phase-shift oscillator

2. To provide positive feedback (360° of phase shift), the op amp supplies 180°—so each RC network must supply 60°. To determine the frequency that results in a 60° phase shift, we note the following:

    $\tan 60° = 1.73 = X_C/R$ where $X_C = 1/(2\pi fC)$

    Using this equation, determine the predicted frequency.

    f = _____

Chapter 8 Oscillators   73

3. Run PSpice and display the output waveform from 0 to 150ms.

   (a) Record below the experimental frequency of oscillation:

   Frequency of oscillation = _____

   (b) Is the above frequency of oscillation *approximately* equal (within 50%) to the calculated value of step 2?

   Yes   No

   (c) Why does the amplitude increase during the early cycles?

   (d) What limits the steady-state amplitude?

4. Referring to Figure 8.1, display the waveforms at points 1, 2, and 3. (Be sure to label the wire segments 1, 2, and 3 as shown.)

   (a) Do the waveforms show the 60° phase shift? (Suggestion: Shift the X-axis to the 100ms-150ms range.)

   Yes   No

   (b) Does the amplitude drop as we move from point 1 to point 3?

   Yes   No

5. Using Fourier analysis (**Plot, X-Axis Settings, Fourier**), generate the frequency spectrum of Figure 8.2.

   (a) Is there a dominant fundamental frequency?

   Yes   No

   (b) Does the fundamental frequency approximately equal the time-domain frequency measured in step 3?

   Yes   No

Colpitts

6. Draw the *Colpitts oscillator* of Figure 8.3, a good choice for higher (radio) frequencies.

PSpice for Windows

74   Part I   Feedback

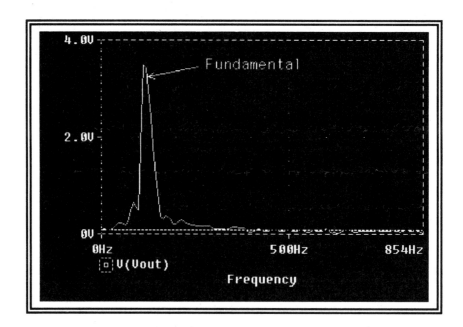

FIGURE 8.2
Phase-shift oscillator frequency spectrum

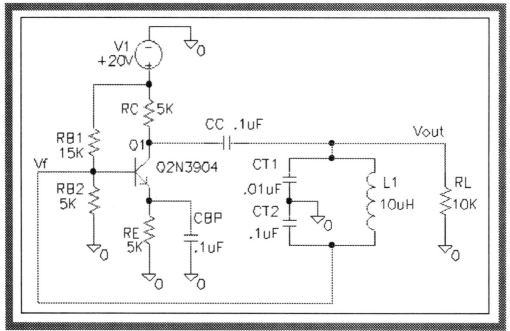

FIGURE 8.3
The Colpitts oscillator

PSpice for Windows

7. Using the equation below, determine the frequency of the tank circuit—which sets the oscillation frequency.

$$F = \frac{1}{2\pi (L \times C_t)^{1/2}} = \underline{\hspace{2cm}} \quad \text{where } C_t = \frac{CT1 \times CT2}{CT1 + CT2}$$

8. However, when we display the output waveform (from 0 to 10µs), there is no signal! The problem is one of insufficient spark. One solution is to pre-charge one of the tank capacitors.

    Using either CT1 or CT2, initialize either capacitor with a small voltage (such as .1V).

9. Again, display the output waveform from 0 to 10µs. This time the signal exists—but clearly, it has not reached steady-state conditions by 10µs.

    Using the *No-Print Delay* option, display the waveform from 200 to 210µs. (See PSpice Note 3.1 from Volume I.)

10. Measure the resonant frequency and compare to the calculated value of step 7. Are they similar?

    f_r (PSpice) = _____

11. Add a plot of Vf (the feedback signal shown on Figure 8.2). Is Vf 180° out of phase with Vout?

    Yes   **No**

12. Generate a frequency spectrum for the Colpitts oscillator (similar to Figure 8.2 for the phase-shift oscillator).

    (a) Is there a DC component? (Does the time-domain signal show a DC component?)

    Yes   **No**

    (b) Is there less distortion (a "sharper" fundamental frequency with fewer harmonics) than with the phase-shift oscillator?

    Yes   **No**

    (c) Does the fundamental frequency approximately equal the time-domain frequency measured in step 10?

    Yes   **No**

Square wave

13. Draw the square-wave oscillator of Figure 8.4, good for audio frequencies.

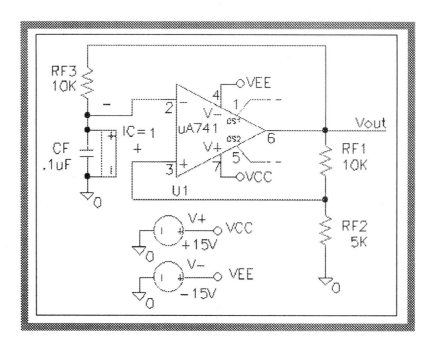

FIGURE 8.4

Square-wave oscillator

14. To predict the oscillation frequency we can reason as follows: The signal at the inverting input oscillates between -5V and +5V. Analysis of an RC pulsed waveform tells us that this requires approximately .7RC time constants. There are two RC swings per cycle. Therefore:

    f = 1/(2 × .7 × RC) = 1/(2 × .7 × 10K × .1µF) = _714Hz_

15. Generate the output waveform using PSpice (0 to 4ms) and measure the oscillation frequency. How does the result compare to the predicted value of step 14?

    f (measured) = _____   f(calculated) = 714Hz

16. By adding waveforms at the V(+) and V(-) points, generate the graph of Figure 8.5. On the graph identify (circle) several trigger points (when the waveform "flips).

PSpice for Windows

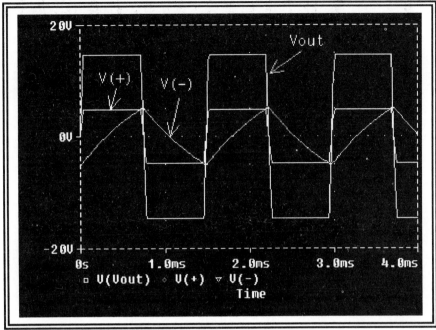

FIGURE 8.5

Square-wave waveforms

Advanced activities

17. By assigning tolerances to the three components that make up the tank circuit of the Colpitts oscillator of Figure 8.3 (CT1, CT2, and L1), generate the histogram of Figure 8.6. What is the percent variation in period between the 10th and 90th %tile?

> Hints:
> - Substitute "break" parts for CT1, CT2, and L1.
> - Assign tolerances by: **CLICKL** on part, **Model**, **Edit Instance Model**, enter tolerance values (such as DEV/GAUSS = 10%), **OK**.
> - Enter the following goal function into *probe.gf* (same file as schematic). (#2# requires that two data points on each side of the peak have lower Y-axis values.)
>
>   period(1) = x2 - x1
>   {
>   1|
>   search forward#2#peak!1
>   search forward#2#peak!2
>   ;
>   }
>
> - Set up the sweep for Transient (200µs - 210µs) and Monte Carlo analysis.
> - When the probe graph appears, switch to *Performance Analysis*.

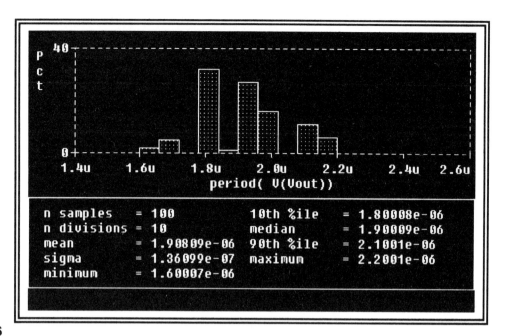

FIGURE 8.6

Period histogram of Colpitts oscillator

18. Design a Colpitts oscillator using an FET instead of a bipolar transistor.

EXERCISES

* By cascading a square-wave oscillator with an op amp integrator, design and test a triangle-wave generator.

QUESTIONS & PROBLEMS

1. What are the three conditions necessary for oscillation?

2. During what period of oscillator operation is the closed-loop gain greater than one? (Hint: how does the signal initially build up?)

3. For a "real-world" oscillator, what serves as the "spark"?

4. For the Colpitts oscillator of Figure 8.3, explain how the loop phase-shift is 360°.

CHAPTER 9

Filters
Passive and Active

OBJECTIVES

* To contrast passive and active filters.
* To compare low-pass, high-pass, and bandpass filters.

DISCUSSION

A *filter* is a device in which the gain is designed to be a special function of the frequency. The major classifications of filters are:

* Low-pass — passes low frequencies.
* High-pass — passes high frequencies.
* Bandpass — passes frequencies within a band.
* Band-stop (notch) — blocks frequencies within a band.

Any of these filters can be either passive or active. A passive filter uses only resistors, capacitors, and inductors. An active filter usually is implemented with an op amp that employs both positive and negative feedback. A *first-order* filter contains a single RC or RL network and rolls off at 20dB/dec; a *second-order* filter contains two RC or RL networks and rolls off at 40dB/dec.

The passive filter

A passive second-order low-pass filter is shown in Figure 9.1. Complex mathematical analysis reveals that it breaks at $1/(2\pi RC)$ and rolls off at 40dB/dec.

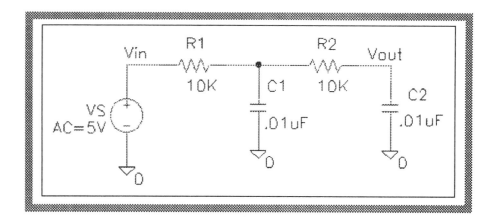

FIGURE 9.1

Passive second-order low-pass filter

An active filter

There are many variations of active filters. One of the most popular is the VCVS (*voltage-controlled-voltage-source*) *Sailen-Key network* configuration of Figure 9.2—popular because R and C network components are of equal value. Complex mathematical analysis yields parameters that are the same as the passive case: a break frequency of $1/2\pi RC$ and a rolloff of 40dB/Dec.

One advantage of using an active filter is that its characteristics depend on gain. For example, changing the gain of the circuit of Figure 9.2 changes its response as shown in Table 9.1. Of the three types of filters, the *Butterworth* is the most common because its response is the smoothest (called "maximally flat").

Response	RF1/RF2	Breakpoint	Gain
Bessel	0.288	1.27fc	1.27 (2.1dB)
Butterworth	0.588	1.00fc	1.588 (4.0dB)
1dB Chebyshev	0.955	0.863fc	1.955 (5.8dB)
2dB Chebyshev	1.105	0.852fc	2.105 (6.5dB)
3dB Chebyshev	1.233	0.841fc	2.233 (7.0dB)

TABLE 9.1

Major filter types

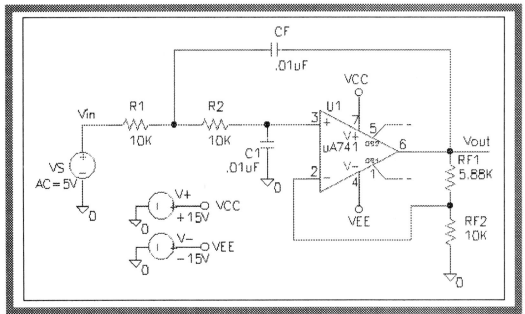

FIGURE 9.2

Active second-order low-pass filter

SIMULATION PRACTICE

1. Draw the second-order, passive, low-pass filter of Figure 9.1, and generate a Bode plot from 1Hz to 10MEGHz. Sketch and label your result on the graph of Figure 9.3. (Reminder: Y-axis is dB(V(Vout)/V(Vin)).)

2. Repeat step 1 for the Butterworth active filter of Figure 9.2 and add the results to Figure 9.3. Be sure to clearly label each curve.

3. Based on Figure 9.3, summarize in the box below the differences between the passive and active filters. (Pay special attention to the frequencies about the breakpoint.)

PSpice for Windows

84   Part I   Feedback

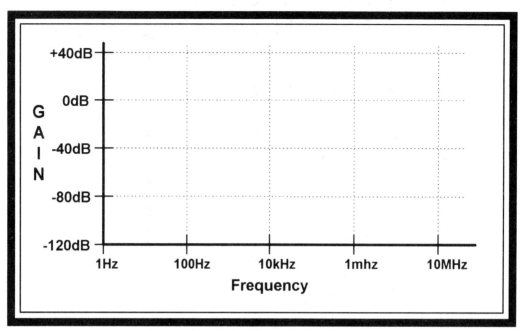

FIGURE 9.3
Bode plot

4. Sweep RF1 over the values shown in Table 9.1 (discussion) and display curves for all five types of filters. Add the four new plots to Figure 9.3 and label all curves.

5. In the box below, summarize the differences between the various types of filter responses (*Bessel, Butterworth,* and *Chebyshev*). (Again, pay special attention to the frequencies about the breakpoint.)

PSpice for Windows

Advanced activities

6. Draw the *multiple-feedback bandpass* filter of Figure 9.4.

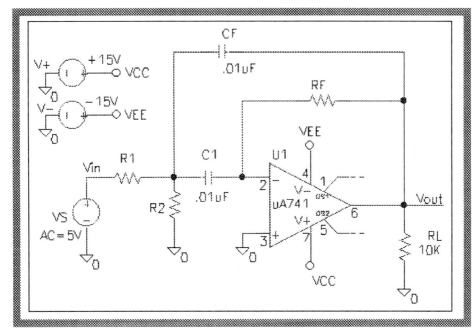

FIGURE 9.4
Bandpass filter

7. When C1 = CF, the formulas listed below determine the values of the resistors. Solve the equations for a center frequency (f_o) of 1kHz, a Q of 5, and a gain (A_o) of 1. (Assume C1 or CF = C = .01μF, and round off all Rs to generally available values.)

   R1 = $Q / (2\pi f_o C A_o)$ = _____

   R2 = $Q / (2\pi f_o C(2Q^2 - A_o))$ = _____

   RF = 2R1

8. Enter the calculated values into the circuit of Figure 9.4 and display the Bode plot. Do the displayed values of A_o, f_o, and Q (f_o/bandwidth) *approximately* match the calculated values?

       Yes   No

9. By reversing the Rs and Cs within the RC networks of Figures 9.1, 9.2, and 9.4 design and test passive and active high-pass and band-stop filters of your choice.

PSpice for Windows

EXERCISES

- Using the active bandpass filter of Figure 9.4, design a circuit to pass the note of A (880Hz) but reject by at least 10dB the next note below (G# @ 830.61) and the next note above (A# @ 932.33).

QUESTIONS & PROBLEMS

1. What are several advantages of using an active filter over a passive filter?

2. Why don't filters generally make use of inductors?

3. What "order" is the bandpass filter of Figure 9.4? (Hint: how many RC networks does the filter have?)

4. Show how a combined (cascaded) low-pass and high-pass filter can be used to construct a bandpass filter.

5. Given a summing amplifier, an inverter, and a bandpass filter, construct a band-stop filter.

CHAPTER 10

The Instrumentation Amplifier
Precision CMRR

OBJECTIVES

- To analyze the instrumentation amplifier.
- To compare the CMRR of an instrumentation amplifier with a single op amp.

DISCUSSION

Suppose we wish to design a bio-feedback circuit. We begin by attaching two electrodes to our arm a slight distance apart. We then send the amplified voltage to a meter, and we close the loop by attempting to mentally control the voltage.

However, this application confronts us with several technical problems. First of all, the bio-feedback voltage is a very small *differential* signal, but our entire arm is immersed in a strong "sea" of electromagnetic interference. Secondly, the transducers attached to our arm supply only very small currents. For the feedback process to work, we must have a high impedance amplifier that rejects the *common-mode* interference signal and amplifies the *differential-mode* bio-feedback signal.

In purely technical terms, we require a high input-impedance differential amplifier of exceptionally high CMRR (common-mode rejection ratio). Also, to match the circuit to different individuals, it would be a real plus if we could easily and quickly control the gain. Such a circuit is called an *instrumentation amplifier* and is shown in Figure 10.1.

PSpice for Windows

Careful mathematical analysis shows that the differential mode (DM) voltage gain of the circuit is:

$$A(DM) = \frac{2R1 + RG}{RG}$$

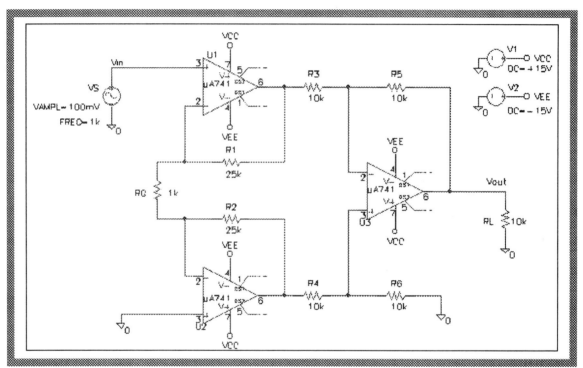

FIGURE 10.1
The instrumentation amplifier

SIMULATION PRACTICE

1. Draw the circuit of Figure 10.1 and set the attributes as shown.

2. Using both calculations and PSpice, determine the differential-mode (DM) voltage gain and place your answers in the table below.

	Calculations	PSpice
V(DM)	_____	_____

3. Using both calculations and PSpice, determine V(DM) when RG = zero and RG = infinity and place your answers in the table given.

	Calculations	PSpice
V(DM)@RG = 0	_____	_____
V(DM)@RG = ∞	_____	_____

4. With RG returned to 1kΩ, determine $Z_{IN}(DM)$.

   Zin(DM) = _____

5. Tie together the two inputs and determine the common-mode voltage gain V(CM). By combining V(DM) and V(CM), determine the CMRR.

   CMRR = _____

6. Based on the results of steps 4 and 5, is the $Z_{IN}(DM)$ and CMRR much higher for an instrumentation amplifier than conventional single op amp amplifiers?

   Yes   No

7. Generate a Bode plot of the instrumentation amplifier and report the bandwidth:

   Bandwidth = _____

   Is the instrumentation amplifier a DC amplifier?

   Yes   No

Advanced activities

8. Remove the ground from the right side of R6 and substitute a +5V source. What effect does this voltage have on Vout?

9. Generate a gain-bandwidth product family of curves for the instrumentation amplifier. Is the gain-bandwidth product a constant?

10. Set resistor tolerances and perform a worst case analysis. Is an instrumentation amplifier more stable than a conventional op amp amplifier?

PSpice for Windows

EXERCISES

- Assuming a 1μV bio-feedback signal, and a .1V common-mode interference signal, test the effectiveness of the instrumentation amplifier as a bio-feedback device. (Suggestion: Select different frequencies for the bio-feedback and interference signals so they can be distinguished in the output signal.)

QUESTIONS & PROBLEMS

1. By adjusting RG, what is the voltage gain range of the instrumentation amplifier of Figure 10.1?

2. What makes the instrumentation amplifier a DC amplifier?

3. For the amplifier of Figure 10.1, what is the voltage gain of the second stage (involving resistors R2 through R6 and op amp U3)?

4. Do transducers generally have a high output impedance? Based on your answer, should the Z_{IN} of the amplifier receiving the transducer's signal be high or low?

5. By adding a Class B transistor buffer to the instrumentation amplifier of Figure 10.1, show how to reduce its output impedance and increase its short-circuit output current.

PART II
Digital

In Part II we switch to digital, where information is stored and transferred in two-state levels rather than continuous analog signals.

To avoid excessive computations, we will find that special techniques are required to simulate digital circuits and to interface analog to digital.

CHAPTER 11
TTL Logic Gates
Analog/digital Interfaces

OBJECTIVES

* To determine the specifications of the TTL logic family.
* To determine how PSpice interfaces analog and digital components.

DISCUSSION

The oldest widely used logic family is TTL (transistor-transistor-logic). In the ideal case, they are simply devices that carry out Boolean algebra. In the real world, however, they have threshold voltages, propagation delays, loading factors, and other considerations of which we should be aware. These specifications are the subject of this chapter.

TTL gates

Figure 11.1 shows that all six basic TTL gates are available to us when using the evaluation version of PSpice.

Analog-to-digital interfacing

In general, PSpice allows us to combine analog and digital devices in any way we wish. In most cases, it is not necessary for the designer to know the "behind-the-scenes" details concerning such mixed circuits. However, when displaying node values under *Probe*, more detailed knowledge of how analog and digital devices are combined may prove essential.

PSpice for Windows

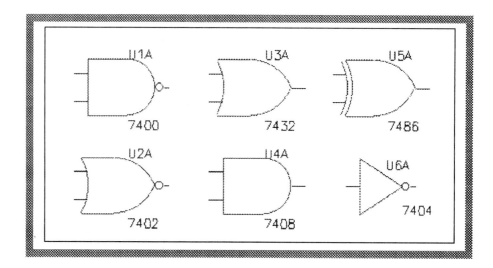

FIGURE 11.1

The basic TTL logic gates

Analog, digital, and interface nodes

PSpice evaluates three types of nodes: *analog*, *digital*, and *interface*. If all the devices connected to a node are analog, then the node is analog. If all the devices are digital, then the node is digital. If both analog and digital devices drive the node, then it is an *interface* node. For each type of node, certain *values* are assigned:

* For analog nodes, the values are voltages and currents.

* For digital nodes, the values are "states," which are calculated from the input/output model for the device, the logic level of the node (0 or 1), and from the output "strengths" of the devices driving the node. Each "strength" is one of 64 ranges of impedance values between the default values of 2 and 20kΩ. To find the strength of a digital output, PSpice uses the logic level and the values of parameters DRVH (high-level driving resistance) and DRVL (low-level driving resistance) from the device's I/O models.

* For interface nodes, the values are both analog voltages/currents and digital states. This is because PSpice automatically inserts an AtoD or DtoA *interface subcircuit* (complete with power supply) at all interface nodes. These subcircuits handle the translation between analog voltages/currents and digital states.

PSpice for Windows

Figure 11.2 shows a simple mixed analog/digital circuit. Although this is the way the circuit is drawn under *Schematics*, the interface nodes do not show the "invisible" circuitry automatically added by PSpice. For that we turn to Figure 11.3, where we have simply drawn in the interface blocks.

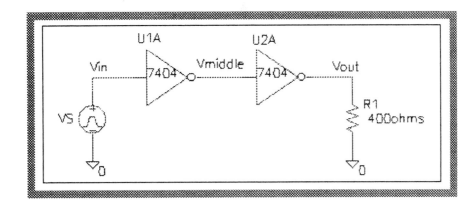

FIGURE 11.2

Mixed component circuit

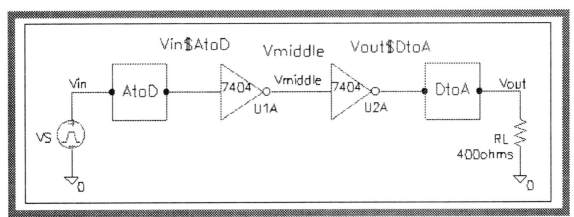

FIGURE 11.3

Circuit showing "invisible" nodes and interface devices

From Figure 11.3, we note the two interface circuits (AtoD and DtoA) created automatically by PSpice. With the wire segments labeled as shown, there are two analog voltage nodes [V(Vin) and V(Vout)] and three digital voltage nodes (Vin$AtoD, Vmiddle, and Vout$DtoA). As we will see, Probe displays digital values quite differently from analog values.

Although the interface circuits do not appear in the schematic, their characteristics are detailed in the output file—as listed below. (Note that PSpice also automatically creates a single interface power supply and ground for the AtoD and DtoA subcircuits.)

**** Generated AtoD and DtoA Interfaces ****

<u>Analog/Digital interface for node Vin</u>
Moving X_U1A.U1:IN1 from analog node Vin to new digital node Vin$AtoD
X$Vin_AtoD1 Vin Vin$AtoD $G_DPWR $G_DGND AtoD_STD
PARAMS: CAPACITANCE= 0

<u>Analog/Digital interface for node Vout</u>
Moving X_U2A.U1:OUT1 from analog node Vout to new digital node Vout$DtoA
X$Vout_DtoA1 Vout$DtoA Vout $G_DPWR $G_DGND DtoA_STD
PARAMS: DRVH= 96.4   DRVL= 104   CAPACITANCE= 0

<u>Analog/Digital interface power supply subcircuits</u>
X$DIGIFPWR 0 DIGIFPWR

.END ;(end of AtoD and DtoA interfaces)

In this chapter we will use several analog/digital circuits to study the basic characteristics and specifications of analog, digital, and interface nodes. Many of our results can be compared to the spec sheet values of Appendix A.

SIMULATION PRACTICE

1. Draw the test circuit of Figure 11.4 and set the attributes as shown.

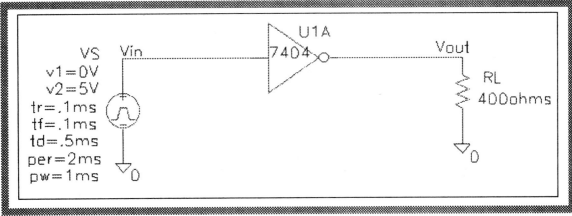

FIGURE 11.4

Analog/digital test circuit

2. Set the transient mode from 0 to 2ms, run a simulation, generate the default Probe graph, and bring up the *Add Traces* dialog box (**Trace, Add**).

   (a) Enable **Analog**, **Voltages**, and **Currents**. Note the following available analog traces:

   Time, V(Vin), V(Vout), V($G_DGND), V($G_DPWR), I(RL), I(VS), V(VS:-).

   (b) Disable **Analog** and enable **Digital**. Note the following available digital traces:

   Vin$AtoD, Vout$DtoA.

   (c) For both the analog and digital cases, enable and disable **Alias Names** and note the additional (alternative) names of the analog nodes.

   (d) On the test circuit of Figure 11.4, sketch in an AtoD and DtoA converter (see Figure 11.3). From the list of (a) and (b), identify and label the analog and digital voltage nodes.

3. Using *analog* trace names, generate the input/output *analog* curves of Figure 11.5.

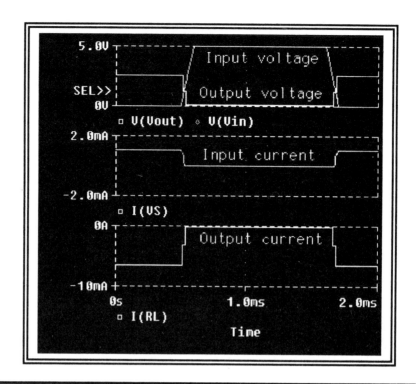

FIGURE 11.5
Threshold test results

4. Based on the results of Figure 11.5, fill in the following voltages and currents in the tables provided and compare to the 7404 specifications given:

   * **Input HIGH and LOW voltages (V_{IH} and V_{IL})** — *the minimum and maximum input voltage that is guaranteed to represent the HIGH and LOW states.* (What values of V_{IN} will give a V_{OUT} of .4V and 2.4V?) :

	V_{IH}	V_{IL}
PSpice		
Spec Sheet	2.0V (min)	0.8V (max)

PSpice for Windows

- **Output HIGH and LOW voltages (V_{OH} and V_{OL})** — *minimum and maximum logic 0 and 1 output voltages.* (What is V_{out} when V_{in} is 0V and +5V?)

	V_{OH}	V_{OL}
PSpice		
Spec Sheet	2.4V (min)	0.4V (max)

- **Input HIGH and LOW currents (I_{IH} and I_{IL})** — *maximum input currents at the HIGH and LOW input states.* (What is I_{in} when V_{in} is 2.4V and .4V?)

	I_{IH}	I_{IL}
PSpice		
Spec Sheet	40μA	1.6mA

- **Output HIGH and LOW currents (I_{OH} and I_{OL})** -- *typical logic 1 and 0 output currents.* (What is I_{OUT} when V_{OUT} is maximum and minimum?)

	I_{OH}	I_{OL}
PSpice		
Spec Sheet	16mA	400μA

Output short circuit current (I_{OS})

5. Short the output (reduce RL to .001ohms) and generate the output *analog* current plot of Figure 11.6.

    Based on the results, fill in the output short circuit current for the output LOW state (V_{in} = 5V) in the table below.

	I_{OS}
PSpice	
Spec Sheet	25mA

PSpice for Windows

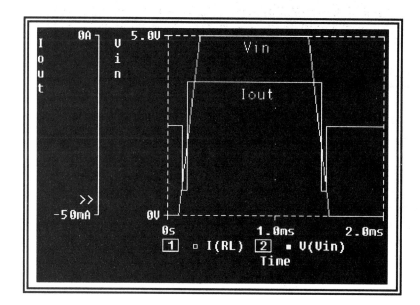

FIGURE 11.6
Output short circuit current

Propagation delay

6. Draw the modified test circuit of Figure 11.7 and set the attributes as shown.

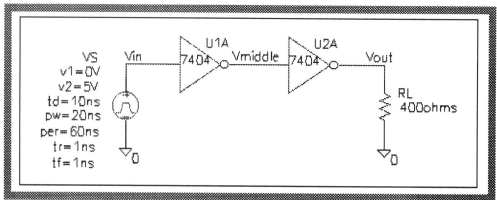

FIGURE 11.7
Propagation delay test circuit

7. Run PSpice and generate the input/output *analog* waveforms of Figure 11.8.

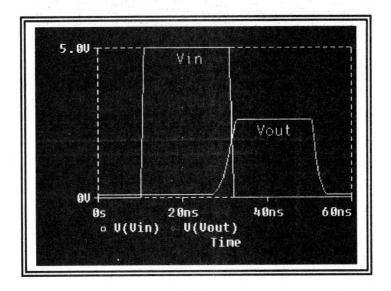

FIGURE 11.8
Propagation delay test results

8. Using any two corresponding points, fill in the value below: (Reminder: The signal has passed through two gates.)

	Propagation Delay
PSpice	_____
Spec Sheet	15ns (typical)

Digital displays

9. To the propagation delay graph of Figure 11.8, add the three digital traces shown in Figure 11.9. (Remember: Enable **Digital** and disable **Analog** in the *Add Traces* dialog box to isolate the digital designators.) Note that all digital signals are listed separately as logic state diagrams.

10. Enable the cursor and set the two traces at arbitrary locations, as shown in Figure 11.10. Observe that the vertical axis of both cursors extends through both the analog and digital sections of the display.

11. Move the cursors back and forth and note that the logic state indicators at the far left of the digital display change accordingly. (0 = low, 1 = high, R = rising, and F = falling).

PSpice for Windows

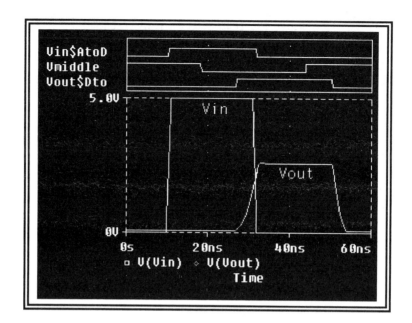

FIGURE 11.9

Adding a purely digital node voltage

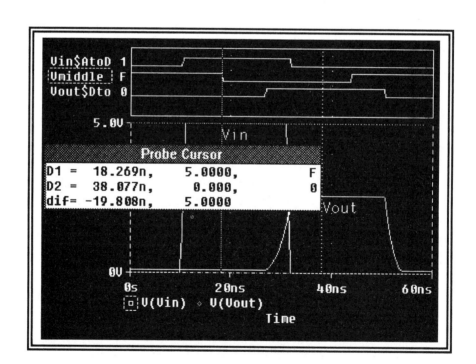

FIGURE 11.10

The cursor in mixed analog/digital displays

PSpice for Windows

Chapter 11 TTL Logic Gates

12. By clicking left (for cursor 1) and right (for cursor 2) on the digital designators at the top of the screen (Vin$AtoD, Vmiddle. etc.), assign cursors 1 and 2 to the top two digital waveforms and note how the corresponding logic state indicators at the far right of the cursor window follow suit as the cursors are moved.

13. Experiment with the cursor until you are familiar with its function in a mixed analog/digital display.

Digital zoom techniques

14. Referring to *Probe Note 11.1,* place "zoom bars" as shown in Figure 11.11 and zoom in on the selected area to generate the waveform of Figure 11.12. Note that both the analog and digital sections are expanded in unison.

    The parallelogram region within the rise and fall times is known as an *ambiguity region* and is the time interval between the earliest and latest time that a transition could occur. (Such timing ambiguities are a major topic of Chapter 14.)

Probe Note 11.1
How do I expand (zoom in on) digital waveforms?

If present, disable the Probe cursor (**Tools, Cursor, Display**).

To place the zoom bars, move the mouse cursor arrow to either side of the desired zoom area, **CLICKLH** to create the first bar, and drag the second bar to the desired location (as shown in Figure 11.11), release the left button, **View, Area**. Note that both the analog and digital regions are zoomed in unison.

By way of review, remember that we can also zoom in on the analog waveforms by **CLICKLH** to drag a square about the area to be zoomed, **View, Area**. Also, we are always free to zoom in on a waveform by changing the X-axis range (**Plot, X-Axis**, etc.).

To return to the original waveforms: **View, Fit**.

108 Part II Digital

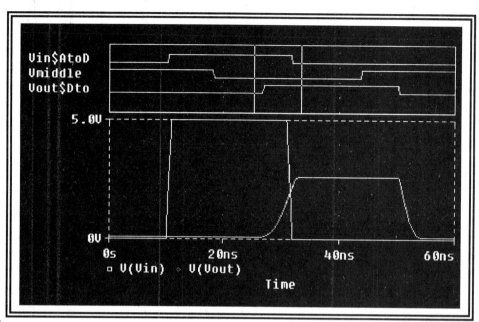

FIGURE 11.11

Placing zoom bars

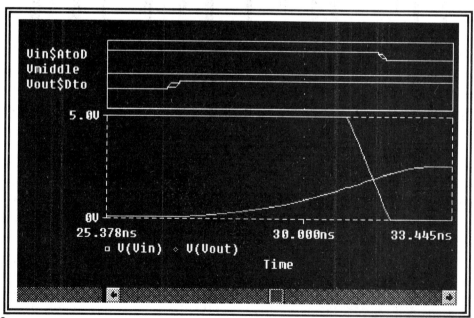

FIGURE 11.12

Expansion to zoom bars

PSpice for Windows

15. Examine *Probe Note 11.2* and change the size of the digital display from 33% of the total waveform space to 50%. (When finished, return to 33%.)

Probe Note 11.2
How do I change the size of the digital display?

To change the size of the digital display in the Probe window: **Plot, Digital Size**, change the *Percentage of Plot to be Digital* from 33 to 50, **OK**.

Markers in a Digital Circuit

16. Place markers on our test circuit as shown in Figure 11.13 and generate the waveforms of Figure 11.14.

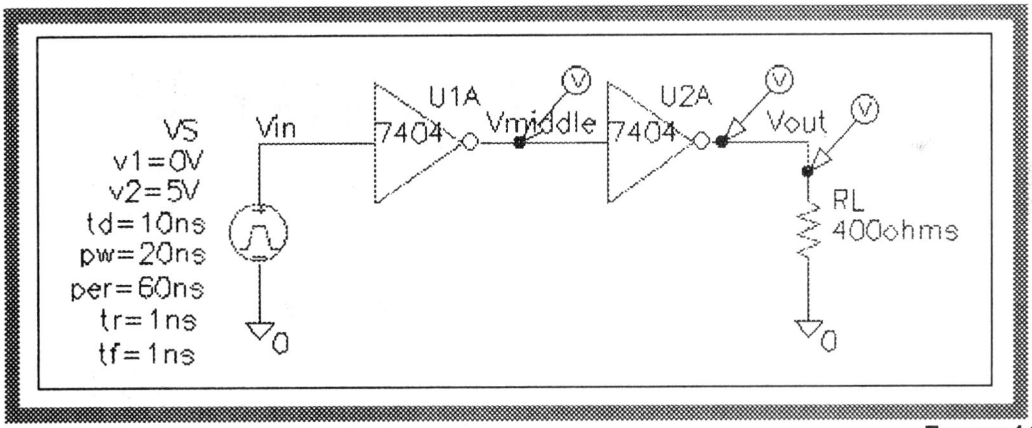

FIGURE 11.13
Placing analog and digital markers

17. Viewing the results (Figure 11.14):

    (a) Are digital waveforms generated only when markers are placed at digital nodes?

        Yes   No

    (b) Are analog waveforms generated when markers are placed at *either* analog or mixed nodes?

        Yes   No

PSpice for Windows

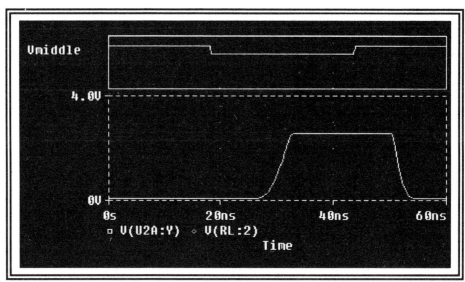

FIGURE 11.14

Marker waveforms

Advanced activities

18. From step 4, we find that the minimum allowed logic 1 voltage level is 2.4V.

    Using the goal function below, set up RL as a parametric variable (Figure 11.4) and generate the graph of Figure 11.15. Based on this graph, what is the minimum allowed value of RL that will guarantee a logic 1 output voltage? Is the value below 400Ω?

    (Note: Be sure to use a text editor for the goal function and store in same directory/file as the schematic.)

    ```
    voutHIGH (1) = y1
    {
       1|
           search forward MAX !1
       ;
    }
    ```

19. Perform a parametric sweep of temperature (suggest -50 to +50 in increments of 25) and display the waveforms. Does temperature influence the waveforms in any significant manner?

PSpice for Windows

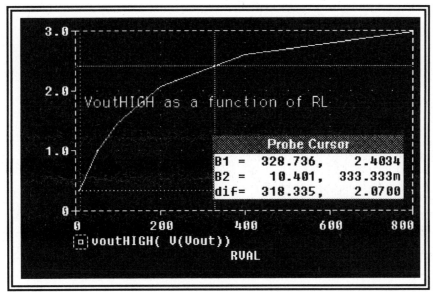

FIGURE 11.15
Performance analysis of VoutHIGH versus RL

20. View traces of the interface power supply (V($G_DGND) and V($G_DPWR)). Are the results as expected?

21. View the model parameters of the 7404. (**CLICKL** to select a 7404, **Edit**, **Model**, **Edit Instance Model**). Is the 7404 a *subcircuit*?

    Yes   No

EXERCISES

- Using a circuit of your design, generate the transfer functions of Figure 11.16 for the conventional 7404 inverter and 7414 Schmitt trigger. (Use analog signals.)
- Test the *clock-edge detection* circuit of Figure 11.17.
- Test the SR flip-flop of Figure 11.18.
- Determine the frequency of oscillation of the circuit of Figure 11.19.

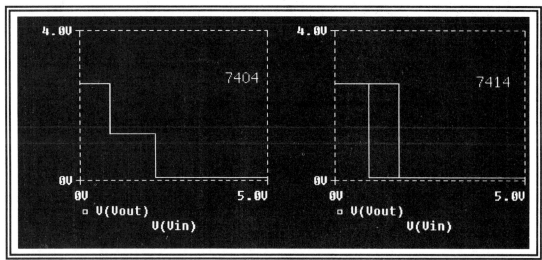

FIGURE 11.16

Transfer function comparison

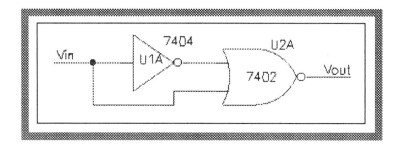

FIGURE 11.17

Clock-edge detection circuit

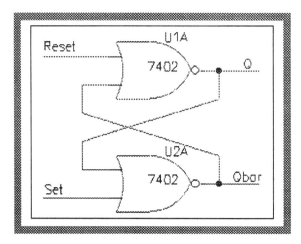

FIGURE 11.18

Set/reset flip-flop

PSpice for Windows

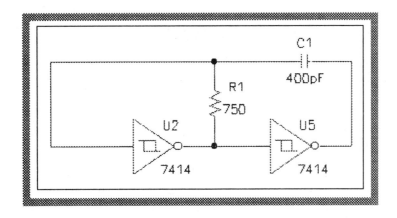

FIGURE 11.19
Oscillator circuit

QUESTIONS & PROBLEMS

1. What are the three possible types of nodes in a mixed analog/digital circuit?

2. Where does PSpice automatically place an AtoD interface circuit? A DtoA interface circuit?

3. Can we measure current at a purely digital node?

4. Regarding the digital gate indicators below, what do the terms "A" and "Y" indicate (to the right of the colon)?

    V(U1A:A)   V(U2A:Y)

PSpice for Windows

5. What is an *ambiguity region*?

6. Explain how the 7414 transfer function of Figure 11.13 shows the effects of *hysteresis*.

CHAPTER 12
Digital Stimulus
File Stimulus

OBJECTIVES

- To generate a variety of digital waveforms using the digital stimulus generator feature of PSpice.
- To generate digital waveforms from information stored in a file.

DISCUSSION

In modern computer systems, it is not unusual for the majority of components and interfaces to be purely digital. Quite often, only the initial input and final output are interfaced to analog devices. Furthermore, it is common for most digital signals to appear in parallel —that is, in groups of 4, 8, 16, or 32.

For these reasons, PSpice added several important features to their simulation software. For routing parallel digital signals, they provided a *bus system*; for generating parallel signals, they provided the special *logic level* and *digital stimulus* devices of Figure 12.1.

The system bus

In a digital system, information is usually stored and moved about in *groups* of bits. To simplify the diagramming of such circuits, Schematics provides the system bus—a single heavy line drawn on the schematic screen and properly labeled. All bus-system wires are simply connected to this single bus line, and all connections identified by labels.

PSpice for Windows

116 Part II Digital

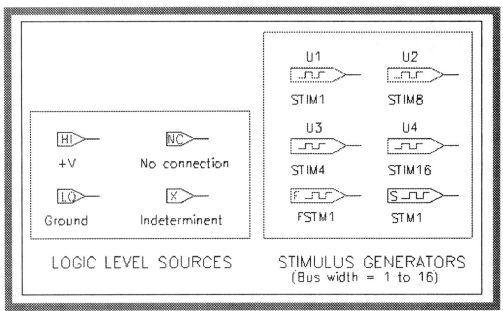

FIGURE 12.1
The logic level sources and digital stimulus devices

Logic level devices

The logic level devices are straightforward and easy to use. They simply apply a constant logic 1 (+5V for TTL) or logic 0 (0V for TTL) to any single node.

For example, the two-input NAND gate of Figure 12.2 uses the HIGH logic level (+5V) source to permanently enable one input and turn the NAND gate into an inverter.

The digital stimulus devices

The digital stimulus devices of Figure 12.1 are designed to interface directly to a bus system, and are far more complex than the simple logic level sources. To begin with, they are divided into three categories:

- The simple stimulus devices (STIM1, STIM4, and STIM16) are programmed for up to 16 parallel and independent digital signals.

PSpice for Windows

- The *stimulus editor (DigStim)* device (STM1) is the digital counterpart to analog's VSTIM, and provides a stimulus editor window to create a variety of waveforms that can be easily modified and selected.

- The *file stimulus (FileStim)* device (FSTM) is used when the number of stimulus commands is very large, and obtains the stream of digital signals from a file.

As always, the best way to learn the use of such sophisticated devices is by way of example.

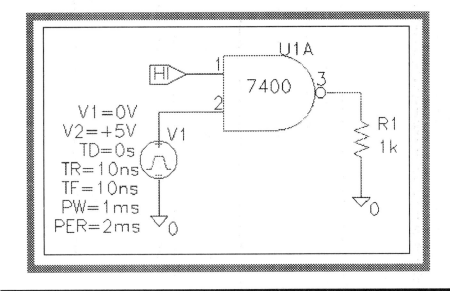

FIGURE 12.2
Use of the logic level source

SIMULATION PRACTICE

Example 1 - Generating a simple clock with a stimulus device

For our first example, we will replace the analog voltage source of Figure 12.2 with a digital stimulus source.

1. Draw the test circuit of Figure 12.3 and set the attributes as shown.
   - STIM1 is from symbol library *source.slb*. (STIM1 feeds a single bus line.)
   - Input and output wire segments are labeled with "stim" and "Vout" for convenient display of waveforms under Probe.

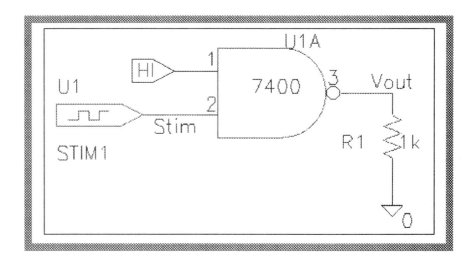

FIGURE 12.3
Replacing analog stimulus with digital stimulus

2. To generate a repeating clock waveform with the same attribute values as V1 in Figure 12.2, **DCLICKL** on device STIM1 to bring up the *Part Name:* dialog box.

   Fill in the parameters as listed in Table 12.1, leaving the default values unchanged. (Be aware that after processing a GOTO statement, the next statement gets executed without delay.)

Chapter 12 Digital Stimulus

ITEM	VALUE	DESCRIPTION
WIDTH =	1	There is one output signal
FORMAT =	1	Output levels specified in binary
DIG_PWR =	$G_DPWR	Default +5V power supply
DIG_GND =	$G_DGND	Default 0V ground
IO_MODEL =	IO_STM	Default IO model
IO_LEVEL =	0	Default interface subcircuit
TIMESTEP =	\<blank\>	Used only if timing specified by clock cycles
COMMAND1 =	+0ms 0	At 0ms, the logic level is binary 0.
COMMAND2 =	LABEL=STARTLOOP	Define a target label.
COMMAND3 =	+1ms 1	1ms later, the output goes to 1
COMMAND4=	+1ms 0	1ms later, the output goes to 0
COMMAND5 =	+1ms GOTO STARTLOOP 3 TIMES	1ms later, branch to STARTLOOP and execute next command (COMMAND3) without delay. Repeat 3 more times for a total of 4.

TABLE 12.1
Example 1
Digital stimulus parameters

3. Run PSpice and generate the digital waveforms of Figure 12.4.

   Are the waveforms as expected? (Remember, after a GOTO statement is processed, the next statement is executed without delay.)

   Yes   No

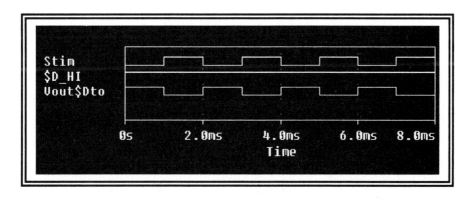

FIGURE 12.4
Example 1 waveform set

Example 2 - Generating two signals with the stimulus editor.

Remaining with the NAND gate of Figure 12.3, our second example will generate both input signals using the special *digital stimulus* source. One input will be a "gating" input, and the other will be a "clock" input. (<u>Reminder</u>: The special digital stimulus source is similar to analog's VSTIM, in that the designer can create a completely custom waveform.)

4. Draw the circuit of Figure 12.5, using part *DigStim* from *source.slb*. (Refer to *Schematics Note 12.1* when setting up the bus system.)

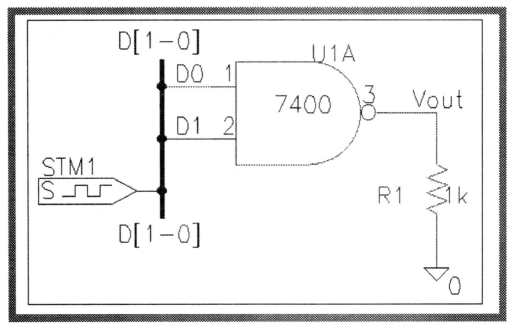

FIGURE 12.5
Parallel digital stimulus and bus system test circuit

5. To test the circuit of Figure 12.5, we will create the input stimulus waveform set of Figure 12.6.

    First, determine the output waveform and draw your prediction in the space provided in Figure 12.6. (Hint: when the inputs are in the HI Z state, expect the output to be in the "X" unknown state.)

PSpice for Windows

Schematics Note 12.1
How do I add a system bus?

To add a system bus, perform the following:: **Draw**, **Bus** to create "pencil", **CLICKL**, **DRAG** to draw a bus segment, **CLICKL** to anchor, **DRAG** to draw next bus segment (if any), repeat until bus is complete, **CLICKR** to abort.

Next, label the bus as follows: **DCLICKL** on any part of the bus to bring up the *Set Attribute Value* dialog box, enter the name of the bus and the numerical bus elements in square brackets. For example, the following name specifies a data bus of two signals (D0 and D1).

$$D[1-0]$$

Next, make all connections to the bus using **Wire** segments.

Finally, label all wire segments with allowed names (using the above example, D0 and D1), making sure that like signals have the same label.

Be aware that a bus can connect to another bus if one is a subset of the other.

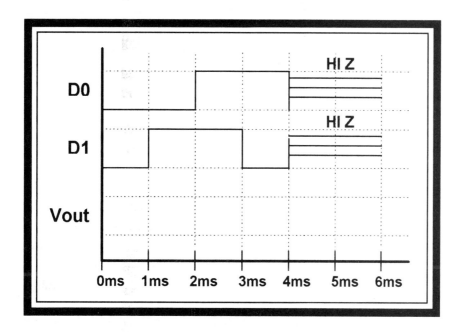

FIGURE 12.6
Stimulus waveform set

6. The next step is to program STM1 to generate the waveform set of Figure 12.6. This we do by following the directions of *Schematics Note 12.2* to generate the equivalent stimulus waveform set of Figure 12.7. (Note: Figures 12.6 and 12.7 are equivalent because hexadecimal 0, 1, 2, and 3 are equal to digital 00, 01, 10, and 11.)

Schematics Note 12.2
How do I use the digital stimulus editor?

To program *DigStim* (STM1) of Figure 12.5 to generate the waveforms of Figure 12.7, perform the following steps:

1. **DCLICKL** on *DigStim* (STM1) to bring up the *Edit Stimulus* dialog box, enter Vtest, **OK** to bring up the *Stimulus Editor* and *Modify STM* dialog box, enter the parameters listed below, **OK**, and note the "unknown" (X) waveform produced.

$$\text{Bus Width} = 2 \text{ bits}$$
$$\text{Time Step} = 1\text{ms}$$
$$\text{Bus Format} = \text{Hex}$$

2. To change the X-axis range: **Plot**, **Axis Settings**, **User Defined**, enter 0 to 6ms, **OK**.

3. From the unknown (X) waveform, we will create the desired waveform set in 1ms segments. To create the segment from 0 to 1ms: **Edit**, **Add Transitions**, **To Value**, enter 00 in *New Value* dialog box, **OK** to create "pencil", **CLICKL** at 0ms, and **CLICKR** to abort. Repeat the process at 1ms (10), 2ms (11), 3ms (01), 4ms (ZZ) to generate the waveform set of Figure 12.7.

4. To save the programming, **File**, **Save**. (Note the *Vtest* on the STM1 output.)

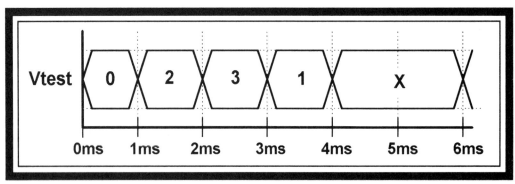

FIGURE 12.7
Stimulus editor waveform

7. Run PSpice and examine signals *D0, D1, Vout$DtoA* (digital) and signal *Vout* (analog).

   (a) Do D0 and D1 match the input waveforms of Figure 12.6?

       Yes   No

   (b) Does the digital output waveform (Vout$DtoA) match your prediction of Figure 12.6?

       Yes   No

   (c) Does the analog output waveform (Vout) match the digital output waveform?

       Yes   No

   (d) Does the unknown (X) output signal (Vout) between 4ms and 6ms lie between .4V (maximum logic 0 output) and 2.4V (minimum logic 1 output)?

       Yes   No

Example 3 - Generate a signal stream with File stimulus

Referring to the first NAND gate circuit of Figure 12.3, our third example will generate the digital signal stream from a file.

8. Draw the circuit of Figure 12.8. (Bring up the circuit of Figure 12.3 and substitute the FSTM1 device for the STIM1 device.)

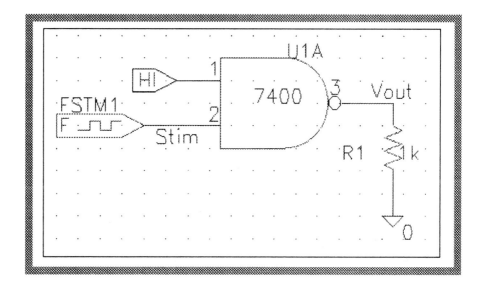

FIGURE 12.8
Replacing digital stimulus with file stimulus

PSpice for Windows

9. *Using a text editor*, open a file (such as NAND.TXT) and enter the signal stream data below: (Be sure to leave the space after STIM.)

   STIM

   0ms 0
   +1ms 1
   +2ms 0
   +1ms 1
   +2ms 0
   +1ms Z

10. Based on the stream of data defined in step 9, predict the output waveform on Figure 12.9.

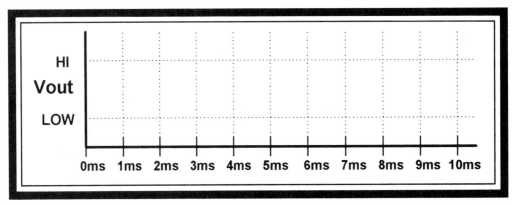

FIGURE 12.9
File stimulus signal stream

11. **DCLICKL** on the file stimulus device (FSTM1), bring up the Part Name dialog box, and enter the data of Table 12.2.

Name	Value	Comment
FileName=	NAND.TXT	Include complete directory path
SigName=	STIM	Signal name
IO_MODEL=	IO_STM	Default value
IO_LEVEL=	0	Default value
ipin[PWR]=	$G_DPWR	Default value
ipin[GND]=	$G_DGND	Default value

TABLE 12.2
FSTM1 Part Name data

PSpice for Windows

Chapter 12 Digital Stimulus

12. Run PSpice and generate the input/output waveforms of Figure 12.10.

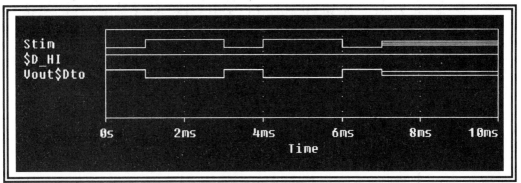

FIGURE 12.10

File stimulus waveforms

13. Referring to Figure 12.10:

    (a) Does the input signal (Stim) match the stream data of step 9?

        Yes   No

    (b) Does the output signal (Vout) match your predictions of Figure 12.9?

        Yes   No

Advanced activities

14. Using the STIM1 device, generate the single digital stream of Figure 12.11.

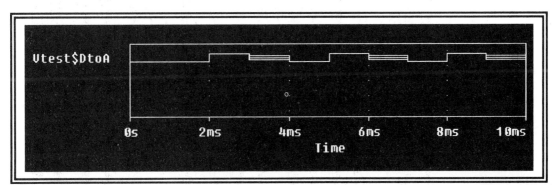

FIGURE 12.11

STIM1 test signal

PSpice for Windows

Exercises

- The 7451 IC of Figure 12.12 consists of two AND and one OR gate internally connected. Using a STIM4 digital stimulus device (INCR mode), determine how the gates are arranged inside the IC block.

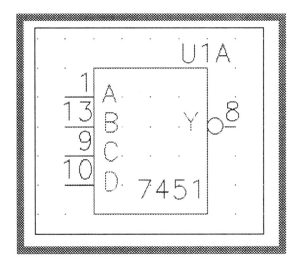

FIGURE 12.12

7451 AND/OR digital IC

QUESTIONS & PROBLEMS

1. How does the use of a bus simplify digital schematics?

2. What is the digital stimulus symbol for each of the following?

   (a) unknown

   (b) High Z

PSpice for Windows

3. What is the purpose of the FORMAT parameter? (Hint: see Table 12.1.)

4. Under what circumstances would the file stimulation device (FSTM) be helpful?

CHAPTER 13
Flip-Flops
Edge versus Level Triggered

OBJECTIVES

- To compare the level-active and edge-trigger D-latch.
- To demonstrate the characteristics of the JK flip-flop.

DISCUSSION

The flip-flop is the basic unit of data storage. As shown by the circuit of Figure 13.1, a single data bit is stored by positive feedback latching action. Table 13.1 shows how the logic 1 (set) and logic 0 (reset) states are entered into the latch.

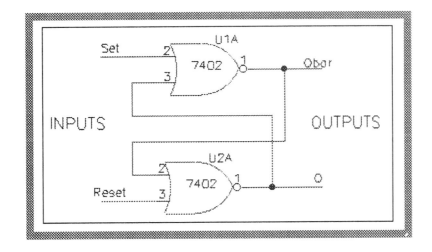

FIGURE 13.1
The SR (set/reset) flip-flop

PSpice for Windows

S	R	Q	Qbar	Comments
0	0	Q	Qbar	No change
1	0	1	0	Set
0	1	0	1	Reset
1	1	0	0	Illegal

TABLE 13.1

The SR flip-flop function table

Synchronous versus Asynchronous

The set (S) and reset (R) inputs to the SR latch of Figure 13.1 are *asynchronous* because they are not clocked and can occur at any time. The more complex flip-flops to follow will also include *synchronous* inputs that must be clocked into the circuit.

The D-Latch

The most basic type of integrated circuit (IC) flip-flop is the D-latch. As shown in Figure 13.2, it comes in two major forms: *level-triggered* (7475) and *edge-triggered* (7474).

- The 7475 level-triggered version offers two flip-flops in a single package. Because the outputs (1Q and 2Q) follow the data inputs (1D and 2D) only when the clock (C) is high, the inputs are synchronous. When the clock is low, the outputs are isolated from the inputs.

- The synchronous input to the 7474 is similar to the 7475, except it is edge-triggered rather than level-triggered. That is, with the 7474, data on the D input is transferred to the Q output on the *low-to-high transition* of the clock pulse.

    The active low *preset* (PRE) and *clear* (CLR) signals are asynchronous because they can be activated at any time. They override the clock input and directly set the Q output to 1 and 0.

    The function table of Table 13.2 compares the synchronous inputs of the 7474 and 7475.

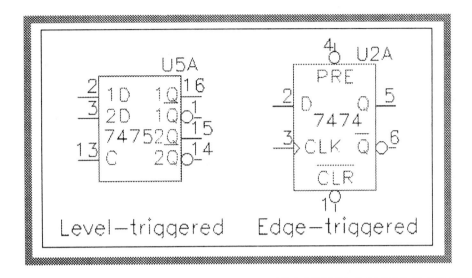

FIGURE 13.2
The D-latches

C	D	Q	Qbar	Comments
1	0	0	1	Q = D = 0
1	1	1	0	Q = D = 1
0	X	Q	Qbar	Data latched

(a) 7475

C	D	Q	Qbar	Comments
0	X	Q	Qbar	No change
1	X	Q	Qbar	No change
↑	1	1	0	Set
↑	0	0	1	Reset

(a) 7475

TABLE 13.2
The D-latch function tables

The JK flip-flop

The JK "master/slave" flip-flop is the most versatile of the three basic types (SR, D, and JK) and offers several advantages over the other two:

- In includes two internal latches, a *master* and a *slave*. This arrangement prevents feedback oscillations by isolating the input from the output.

- In offers a *toggle* mode, in which the outputs *change* state.

The JK flip-flop also comes in *pulsed* (level-active) and *edge-triggered* versions. This section will include only the 7476 pulsed version because it is the only one offered by the present evaluation version of PSpice.

The 7476 pulsed-mode master/slave JK flip-flop is shown in Figure 13.3. It offers both asynchronous (PRE and CLR) and synchronous inputs. As shown by the function table of Table 13.3, synchronous input data is loaded into the master while the clock goes HIGH and transferred to the slave on the HIGH-to-LOW clock transition.

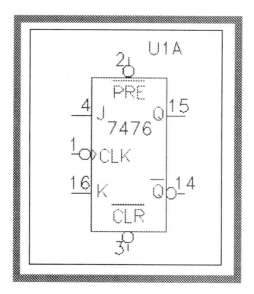

FIGURE 13.3

The 7476 JK flip-flop

CLK	J	K	Action
↑↓	0	0	none
↑↓	1	0	set (Q high)
↑↓	0	1	reset (Q low)
↑↓	1	1	toggle (Q changes)

TABLE 13.3
The JK flip-flop function table

SIMULATION PRACTICE

NOTE: If you experience difficulty with any digital IC (such as only receiving unknowns on outputs), set the initial state to zero (**Analysis**, **Setup**, **Digital Setup**, **All 0**, **OK**, **Close**).

The D-Latch

The 7475 Level-Triggered D-Latch

1. Draw the 7475 D-latch circuit of Figure 13.4 and set the attributes as shown.

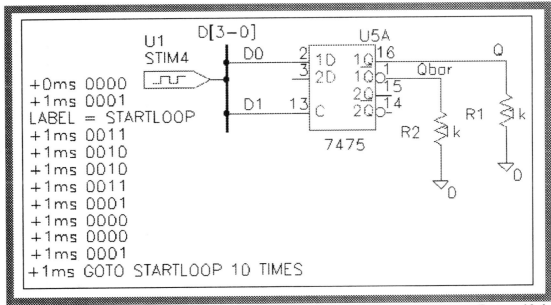

FIGURE 13.4
D-Latch circuit

2. Figure 13.5 shows the input signals to the D-latch. Below the signals, predict (draw) the Q output waveform in the space provided.

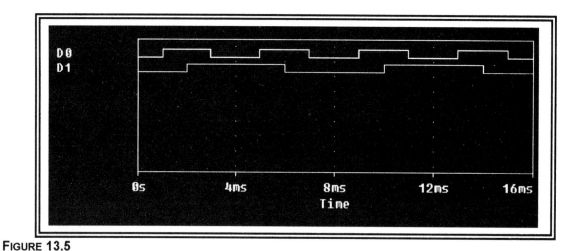

FIGURE 13.5

7475 D-Latch waveforms

3. Generate the Q (Q$DtoA) digital output waveform using PSpice. Were your predictions correct?

   Yes   No

   (Make any necessary corrections to your predicted waveform.)

4. Add the analog output waveform [V(Q)] to your graph, and greatly expand the X-axis about 2ms to generate the graph of Figure 13.6.

5. Viewing the results (Figure 13.6), what causes the delay between D1 and Q$DtoA?

6. Did the digital output (Q$DtoA) go high when the analog output signal [V(Q)] just started its low-to-high transition?

   Yes   No

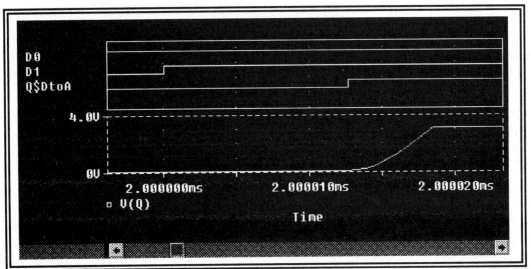

FIGURE 13.6

Expanded risetime waveforms

The 7474 Edge-Triggered D-Latch

7. Substitute the 7474 D-latch for the 7475 of Figure 13.4 and produce the circuit of Figure 13.7.

8. As with the 7475, sketch your predicted Q output waveform on the graph of Figure 13.8.

9. Generate the Q output waveform using PSpice. Were your predictions correct? (Make any necessary corrections to your predicted waveform.)

    Yes   No

10. Comparing the Q output waveforms between the 7474 and 7475, do you see the difference between *level-triggered* and *edge-triggered?*

    Yes   No

PSpice for Windows

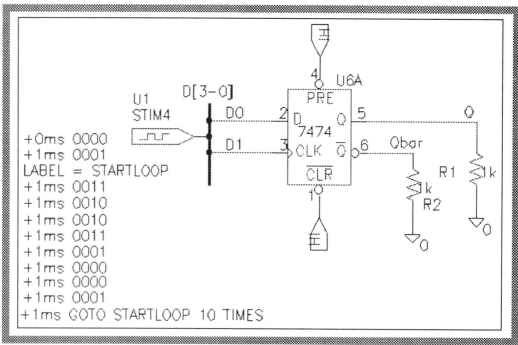

FIGURE 13.7

7474 D-Latch circuit

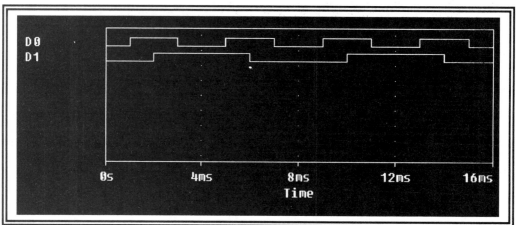

FIGURE 13.8

D-Latch waveforms

PSpice for Windows

The JK flip-flop

11. Draw the test circuit of Figure 13.9 and set the attributes as shown.

12. As before, sketch your predicted Q output waveform on the graph of Figure 13.10.

13. Generate the Q output waveform using PSpice. Were your predictions correct? (Make any necessary corrections to your predicted waveform.)

    Yes   No

14. Reviewing Figure 13.9, what is the frequency relationship between input and output?

    f(out) = _____ × f(in)

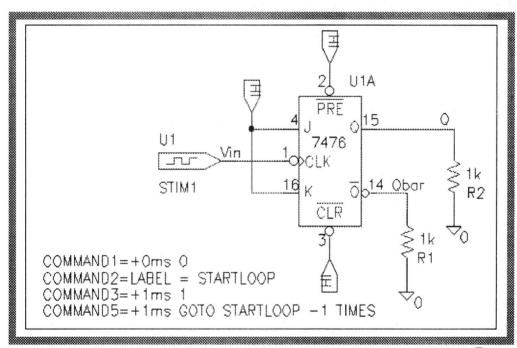

FIGURE 13.9

JK flip-flop programmed for toggle mode

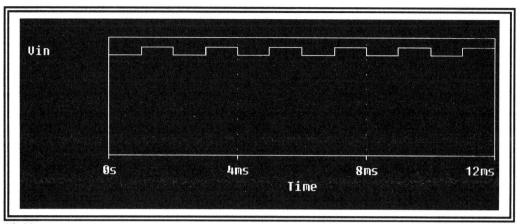

FIGURE 13.10
Flip-flop waveforms

Advanced activities

15. *Using only basic gates*, design and test any or all of the following:

    (a) Level-triggered D flip-flop.
    (b) Edge-triggered D flip-flop. (Hint: make use of the clock-edge detection circuit of Figure 11.17.)
    (c) Pulsed JK flip-flop.

EXERCISES

- Draw and test the parity circuit of Figure 13.11. [As serial data enters the Data input (D1) at the clock rate (D0), the circuit keeps a running total of parity. If an even number of 1s have been received up to a certain time, the Q output will be 0; otherwise, the Q output will be 1.]

- Design, draw, and test a "Jeopardy" circuit (to be used on the popular Jeopardy quiz show), which has three input switches and three output "lights." An answer is revealed by the quizmaster, and the three contestants each try to press a switch first in order to be recognized. Whichever switch is activated first causes the corresponding output to go active—and deactivates the other two switches.

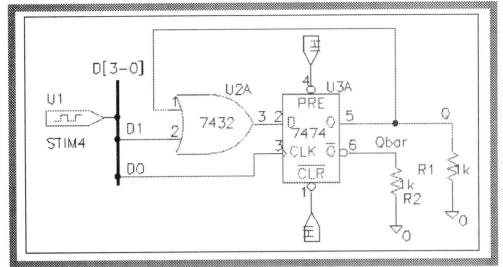

FIGURE 13.11

Parity test circuit

QUESTIONS & PROBLEMS

1. What is the difference between *level-triggered* and *edge-triggered*?

2. With a JK master/slave flip-flop, what happens during the rising clock pulse and the falling clock pulse?

3. Why is the CLR (clear) input called *asynchronous*?

PSpice for Windows

4. By adding an inverter, show how to construct a D-latch from a JK flip-flop.

CHAPTER 14
Digital Worst Case Analysis
Ambiguity and Timing Violations

OBJECTIVES

- To apply worst case analysis techniques to digital circuits.
- To identify and correct a variety of timing violations.
- To analyze the effects of ambiguity.

DISCUSSION

As a production engineer you note something interesting: How can it be that all the circuits before me came off the same assembly line, but 20% of them do not work? As with their analog counterparts, we suspect that the problem is caused by component *tolerances*. From one circuit to the next, there is a varying combination of individual component tolerances, and in 20% of the cases it is causing a problem.

In this chapter we will investigate the following tolerance parameters:

- *Propagation delay* — the time span between an input signal transition and the resulting output response.
- *Setup* — the time span during which a signal must be held steady *before* an action is taken.
- *Width* — the width of a clock pulse.
- *Hold* — the time span during which a signal must be held steady *after* an action is taken.

PSpice for Windows

Ambiguity

As shown in Figure 14.1, PSpice places all digital signals into one of six states:

- 0 for *low*
- 1 for *high*
- R for *rising*
- F for *falling*
- X for *unknown*
- Z for *high impedance*

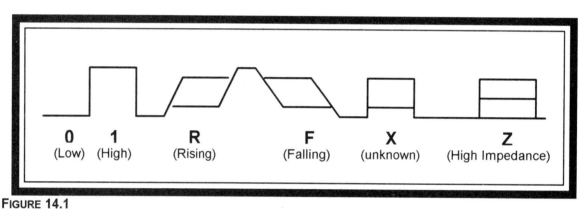

FIGURE 14.1
The six digital states

All R and F regions are known as *ambiguity regions*, because the exact time of transition is not precisely known—all we know for sure is that it falls somewhere in the region. The major source of ambiguity is the variation in a component's *propagation delay*.

For example, Table 14.1 shows propagation delay data for the 7408 AND gate. As shown, the propagation delay is not a constant, but is a range of values between the *minimum* and *maximum* extremes. Most propagation values will of course be clustered about the *typical* value. By definition, the ambiguity (uncertainty) times fall between the minimum and maximum values.

Chapter 14   Digital Worst Case Analysis

	RISING	FALLING
MINIMUM	7ns	5ns
TYPICAL	17ns	12ns
MAXIMUM	27ns	19ns
AMBIGUITY	20ns (27ns – 7ns)	14ns (19ns – 5ns)

TABLE 14.1
7408 Propagation delay summary

Worst case analysis

To handle the problem of tolerance variations in *analog* circuits, we used *worst case* analysis to display the maximum possible variation from the nominal case. To handle the problem of tolerance variations in *digital* circuits, we also use *worst case* analysis to directly display the ambiguity regions.

For example, based on the data of Table 14.1, Figure 14.2 shows the output ambiguity region for just the risetime portion of the input pulse. When the input signal rises at the 40ns point, the output goes high anywhere from 7ns to 27ns later. The 20ns ambiguity region represents the interval between the earliest and the latest time that the low-to-high transition can occur.

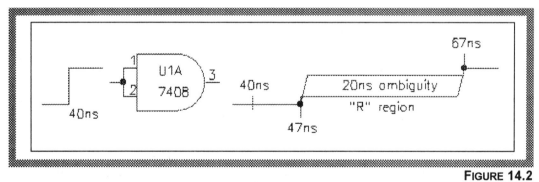

FIGURE 14.2
Risetime ambiguity region

When using worst case analysis to uncover "race" (timing) problems, all possible combinations of propagation delays are generated and *ambiguity regions* are automatically displayed.

PSpice for Windows

SIMULATION PRACTICE

Propagation delay

1. Draw the test circuit of Figure 14.3 and set the digital stimulus as shown. Suggestion: label input/output wire segments *Sin* (signal in) and *Sout* (signal out).

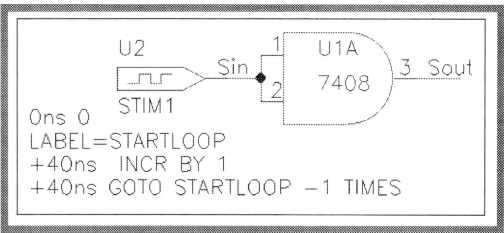

FIGURE 14.3
Propagation delay
test circuit

2. Set up the timing analysis for *minimum* (**Analysis**, **Setup**, **Digital Setup**, **Minimum**, **OK**, **Close**), enable the transient mode, and generate the waveform of Figure 14.4.

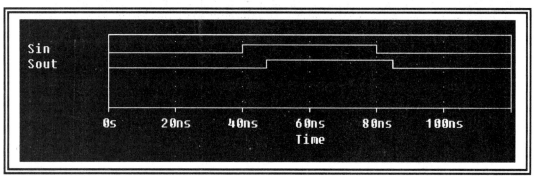

FIGURE 14.4
7408 Minimum
propagation delays

PSpice for Windows

3. Are the *minimum* rising and falling propagation delays the same as shown in Table 14.1?

    Yes   **No**

4. Repeat for the *typical* and *maximum* cases. Do they agree with Table 14.1?

    Yes   **No**

Ambiguity

5. This time, repeat the same process, but using *Worst-Case[Min/Max]* analysis to generate the curves of Figure 14.5.

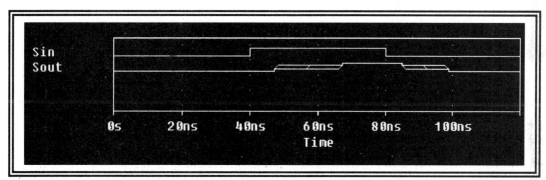

FIGURE 14.5
Worst case run, showing ambiguity regions

6. Examine the results and answer the following:

   (a) Does the risetime ambiguity region match the predictions of Figure 14.2 (and Table 14.1)?

        Yes   **No**

   (b) Does the fall-time ambiguity region match the predictions of Table 14.1?

        Yes   **No**

Convergence hazard

The first hazard we will study is called a *convergence hazard*. It occurs when two or more signals with overlapping ambiguity regions converge at a common point (such as a gate). The overlap period results in a corresponding unknown ("X") output period.

7. Draw the test circuit of Figure 14.6 and set the commands as shown.

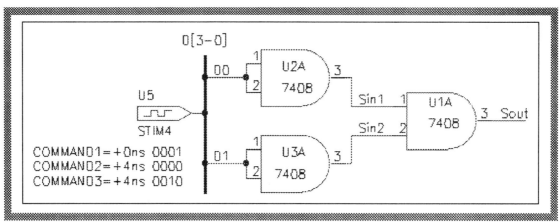

FIGURE 14.6
Test circuit for
CONVERGENCE hazard

8. Figure 14.7 shows the input waveforms to the AND gate array. Assuming ideal AND gates (no propagation delay), draw the expected output waveform in the space provided. (Pencil may be easier to see on the black background.)

9. Using worst-case analysis, run the simulation. Because an error was found by PSpice, **OK** in the *Simulation Errors* dialog box to bring up the *Simulation Message Summary* dialog box. When did the CONVERGENCE hazard occur?

     The CONVERGENCE error occurred @ _____

10. Display the various options available under *Minimum Severity*. In which of the categories listed below did the CONVERGENCE error fall? (Circle your answer.)

     WARNING   SERIOUS   FATAL

PSpice for Windows

Chapter 14 Digital Worst Case Analysis   147

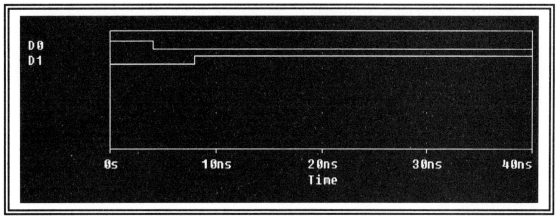

FIGURE 14.7
CONVERGENCE hazard input waveforms

11. Activate the various options available under *Sort By* (except *Section*) and note the changes.

12. **Plot**, **Close** to generate the special "hazard" Probe curves of Figure 14.8.

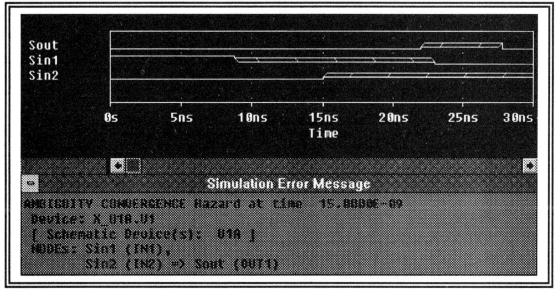

FIGURE 14.8
CONVERGENCE hazard waveform set

PSpice for Windows

13. The CONVERGENCE hazard occurred at 15ns because that is when the input ambiguity regions overlapped. During this overlap period, the output waveform is uncertain.

    (a) What is the state of the output ambiguity region? Circle your answer.

    R   F   X

    (b) From Table 14.1, the minimum rising propagation delay is 7ns. Does the output signal ambiguity occur 7ns after the ambiguity convergence hazard?

    Yes   No

14. Increase the stimulus command times from +5ns to +15ns and rerun PSpice. Do the input ambiguity regions no longer overlap, and was the CONVERGENCE hazard eliminated?

    Yes   No

Cumulative ambiguity hazard

A *cumulative ambiguity hazard* occurs when signals propagate through layers of gates. As the signal passes through each gate, ambiguity accumulates (the ambiguity regions widen). When the rising edge ambiguity overlaps the falling edge ambiguity, an unknown output is created and a hazard is predicted.

15. Draw the test circuit of Figure 14.9 and set the commands as shown.

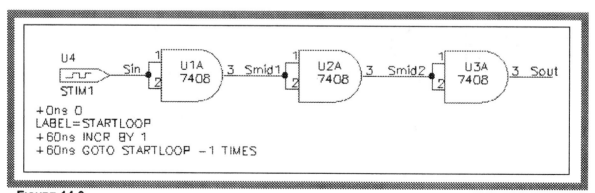

FIGURE 14.9
CUMULATIVE ambiguity test circuit

16. Perform a worst case analysis and generate the hazard curves of Figure 14.10. (Hint: add curves as necessary and expand the X-axis.)

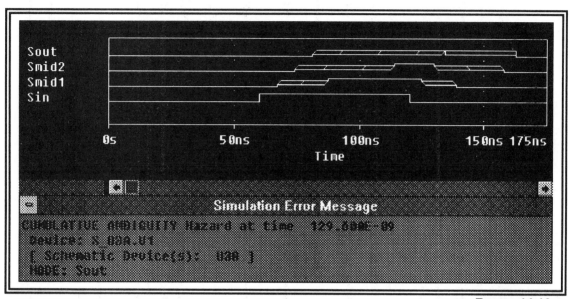

FIGURE 14.10
CUMULATIVE ambiguity waveform set

17. Analyze the curves and answer the following:

   (a) As the input waveform propagates through the gates, does the ambiguity accumulate? (Do the ambiguity regions grow wider?)

       Yes   No

   (b) Is the regular (unambiguous) portion of the signal gradually "squeezed out"?

       Yes   No

   (c) Does the output signal (Sout) show an unknown (X) region where the signal could either be rising or falling?

       Yes   No

18. The CUMULATIVE ambiguity hazard occurred at 129.6ns because that is when PSpice first determined that the rising and falling ambiguity regions would overlap (converge).

    Does the unknown (X) region occur in the output signal 5ns (the minimum falling propagation delay) after the hazard is detected?

    Yes   No

19. Increase the input pulse width from +60ns to +80ns. Did the CUMULATIVE hazard disappear?

    Yes   No

Setup, hold, and width hazards

The *setup*, *hold*, and *width* hazards commonly occur in circuits that are clocked—such as flip-flops. If the width of a clock signal is too low, a *width* hazard is predicted; if a command or data signal is not stable for a sufficient time period *before* a clock occurs, a *setup* hazard is predicted; if a command or data signal is not stable for a sufficient time period *after* a clock occurs, a *hold* hazard is predicted

20. Draw the circuit of Figure 14.11 and set the commands as shown.

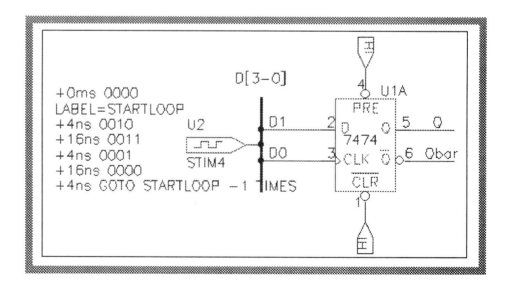

FIGURE 14.11

SETUP, HOLD, and WIDTH test circuit

PSpice for Windows

Chapter 14 Digital Worst Case Analysis

21. Run PSpice and bring up the *Simulation Errors* dialog box, and note the number of errors. **OK** to bring up the *Simulation Message Summary* dialog box and note that the first three hazards are *Setup* (20ns), *Hold* (24ns), and *Width* (40ns).

22. To create three windows corresponding to the first three hazards, **Plot**, **Plot**, **Plot** (Simulation Message Summary), **Close**.

23. First, select the SETUP window of Figure 14.12 (**Window**, select proper window).

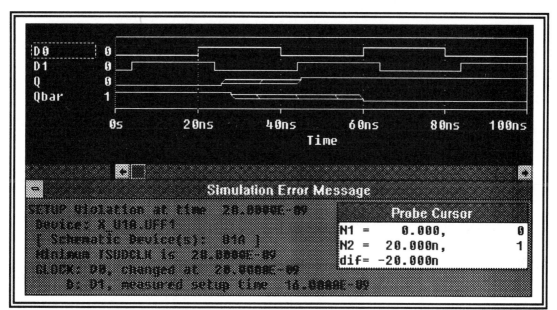

FIGURE 14.12
SETUP violation waveform set

24. The SETUP hazard occurred at 20ns because that is when the system first determined that the time between a data change (D1 at 4ns) and a clock transition (D0 at 20ns) was below the minimum allowed.

   With the help of the *Simulation Error Message* box, determine each of the following:

   Measured SETUP = _____

   Minimum SETUP (TSUDCLK) = _____

   Additional SETUP time required (minimum – measured) = _____

PSpice for Windows

25. Next, select the WIDTH window. The WIDTH hazard occurred at 40ns because that is when the system first determined that the width of the clock signal (D0) was too small. Record each of the following:

    Measured WIDTH = _____

    Minimum WIDTH (TWCLKH) = _____

    Additional WIDTH time required (minimum − measured) = _____

26. Finally, select the HOLD window. The HOLD hazard occurred at 24ns because that is when the system first determined that the data input (D1 at 24ns) changed too quickly *after* the clock (D0 at 20ns) when active high. Record each of the following:

    Measured HOLD = _____

    Minimum HOLD (THDCLK) = _____

    Additional HOLD time required (minimum − measured) = _____

27. Increase the COMMAND times in Figure 14.11 from 4ns/16ns to 8ns/32ns. Were all the violations eliminated?

    Yes   No

Advanced activities

28. Perform a worst case analysis on the oscillator of Figure 14.13 (from Chapter 12) and generate the output graph of Figure 14.14.

    Because of the continuous accumulation of ambiguity in the feedback loop, the output signal quickly reaches the X (unknown) ate and ceases to operate. Yet, we know this does not really ppen in "real life." *We conclude that worst case analysis cannot be used in circuits with asynchronous feedback.*

    To overcome the problem, set all active devices in the feedback loop (the two 7414s) to typical (TYP) by setting model parameter MNTYMXDLY=2 (**DCLICKL** to bring up each *Part Name* dialog box). All other devices, if any, will continue to operate in the worst case mode. Does the circuit now operate properly?

PSpice for Windows

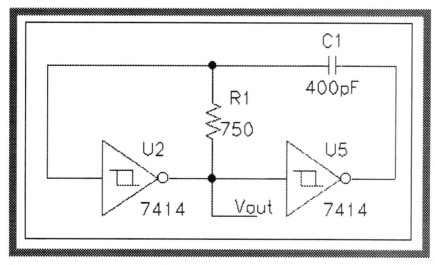

FIGURE 14.13
Digital oscillator

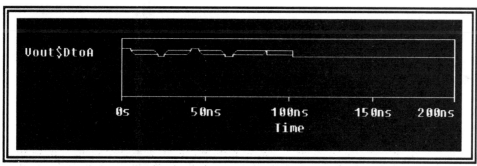

FIGURE 14.14
Ambiguity build-up in oscillator

Exercises

- Using worst case analysis, determine the highest "safe" frequency of operation of the circuit of Figure 14.15.

PSpice for Windows

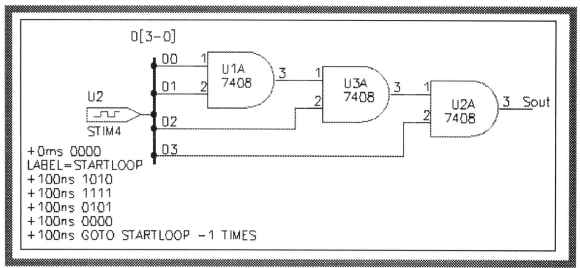

FIGURE 14.15
Exercise circuit

QUESTIONS & PROBLEMS

1. What is the difference between SETUP and HOLD?

2. On a digital waveform, what does the "X" state indicate?

3. Why is it unrealistic to perform a worst case analysis on a feedback loop?

PSpice for Windows

4. The input waveform shown below has input rising and falling ambiguities of 1ns. After passing through the 7404, which adds an additional 3ns (rising) and 2ns (falling) of ambiguity, draw the output waveform (including ambiguity regions).

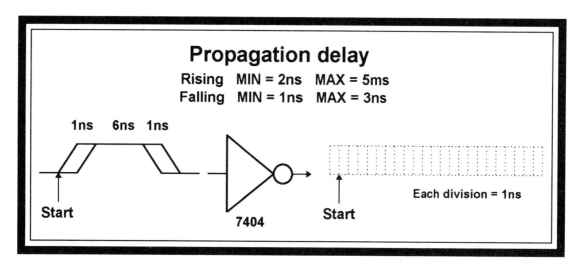

CHAPTER 15

Critical Hazards
Persistence

OBJECTIVES

- To apply worst case analysis techniques to digital circuits.
- To identify and correct a variety of timing violations.
- To analyze the effects of ambiguity.

DISCUSSION

All the violations and hazards of the previous chapter were WARNINGS. As such, they may or may not cause a serious problem. They are simply events that the designer must examine in order to determine if a change in circuit design is warranted. As an example, consider the circuit of Figure 15.1.

When the circuit is processed under PSpice, the waveform set of Figure 15.2 shows that a SETUP WARNING was properly identified at 20ns. This is because the data line transition occurred only 10ns before the active high clock pulse—and a minimum of 20ns is required.

However, this probably will not cause a problem in this case because a second clock pulse comes along at 100ns and properly latches the data. (Note that the ambiguity in the Q signal occurs only after the first clock pulse.) Therefore, the design engineer may very well decide to ignore this particular SETUP WARNING.

PSpice for Windows

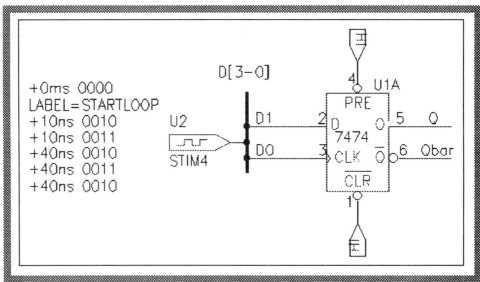

FIGURE 15.1

SETUP test circuit

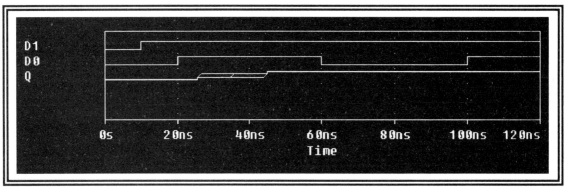

FIGURE 15.2

SETUP warning waveform set

Persistent hazard

A *persistent hazard*, on the other hand, is a timing violation or hazard that can cause an incorrect state to be latched into an internal circuit (such as a flip-flop), or one that is passed on to a primary circuit output.

These more serious persistent hazards, which usually cannot be ignored, are a major subject of this chapter.

Design methodology

Because of violations and hazards, the development of a digital circuit is a two-phase process: *design* and *verification.*

- During *design*, all tolerances are set to typical and we concentrate on the *state response* of the circuit.

- During *verification*, all tolerances are set to worst case and we concentrate on violations and hazards. Starting at a hazard point, we work backwards until the cause of the hazard is uncovered and corrected.

SIMULATION PRACTICE

External ports

1. Draw the CONVERGENCE hazard test circuit of Figure 15.3. Be sure to attach an EXTERNAL_OUT *port* symbol (from library *port.slb*) to the circuit output. (The EXTERNAL_OUT ports represent data outputs to other circuitry.)

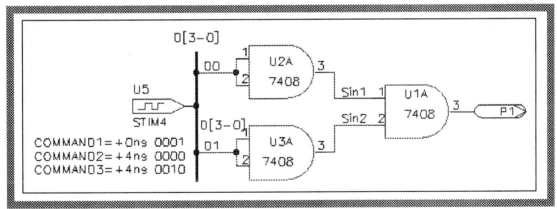

FIGURE 15.3

EXTERNAL_OUT port added to CONVERGENCE circuit

2. Run the simulation and **Plot** (from the *Simulation Message Summary* dialog box) to generate the hazard waveform set of Figure 15.4. As shown two plots are generated:

   - The upper plot displays the persistent hazard at 22ns due to the propagation of a timing hazard to an output port.

   - The lower plot displays the *cause* of the persistent hazard—a convergence hazard at 15ns.

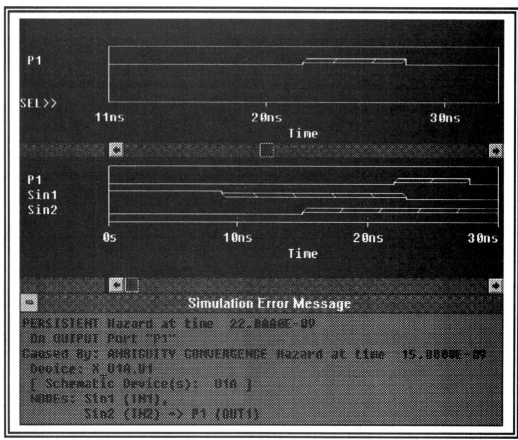

FIGURE 15.4
Persistent hazard and cause

Latched hazard

3. Draw the circuit of Figure 15.5 and set the stimulus commands as shown.

4. Run the simulation and plot the first persistent hazard to generate the hazard waveform set of Figure 15.6.

Chapter 15 Critical Hazards

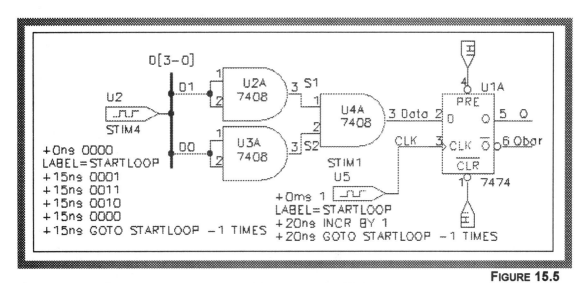

FIGURE 15.5
Internal latch test circuit

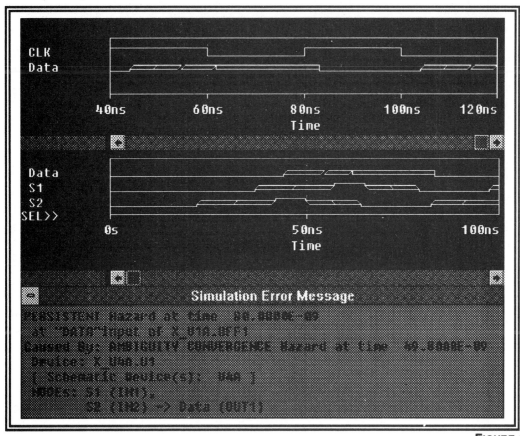

FIGURE 15.6
Persistent hazard and cause

PSpice for Windows

- The upper plot displays a persistent hazard at 80ns because an unknown ambiguity region (state "X") was present at the data input to the flip-flop when the clock went active high. As a result, an unknown value was latched into the flip-flop. This could cause a serious malfunction in the overall circuit.

- The lower plot shows us that the cause of the persistent hazard was the ambiguity convergence hazard at the output of the AND gate array.

Advanced activities

5. Using worst case analysis, determine the maximum frequency of operation of the D-latch-based ripple counter of Figure 15.7. (As the frequency is increased, what is the first hazard that appears?)

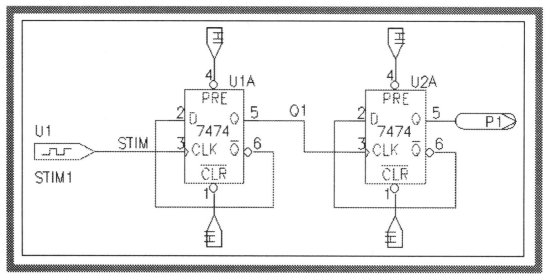

FIGURE 15.7

Worst case test circuit

EXERCISES

Perform the two-step design methodology outlined in the discussion on the circuit of Figure 15.8.

- Set the timing state to *typical*, run the simulation, and generate a waveform set. Do the waveforms indicate that the state response of the circuit is correct?

- Switch to worst case mode and generate a hazard waveform set. Report how you corrected the hazard.

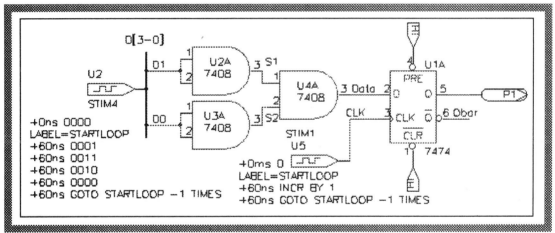

FIGURE 15.8

Design methodology test circuit

QUESTIONS & PROBLEMS

1. Reviewing Figure 15.2, what is the SETUP time of the data (D1) for the *second* clock pulse (D0)?

2. What two basic steps should every design engineer perform on a digital circuit?

3. What two conditions does PSpice look for when tagging a hazard as *persistent*?

4. Does PSpice display both the *cause* and *effect* of a persistent hazard?

CHAPTER 16

Counters
Synchronous versus Asynchronous

OBJECTIVES

- To compare *synchronous* versus *asynchronous* digital circuitry.
- To design and test a variety of counters.

DISCUSSION

A digital counter is a device that keeps track of (counts) clock cycles. Counters are fashioned from a series of flip-flops strung together in sequence, with one flip-flop feeding the next.

The number of different binary states defines the *modulus* (MOD) of the counter. For example, if a counter counts from 0 to 3, it is a MOD-4 counter. A MOD-4 counter must be constructed from 2 flip-flops.

Synchronous versus asynchronous

Digital counters are classified as either *synchronous* or *asynchronous*.

- If the output of one flip-flop is used to clock the next flip-flop, as shown in the MOD-4 counter of Figure 16.1(a), the counter is asynchronous. Asynchronous counters are also called *ripple counters*, because the clock signal ripples from one flip-flop to the next.
- If all the flip-flops are clocked at the same time, as shown in the MOD-4 counter of Figure 16.1(b), the counter is synchronous. Synchronous counters are much faster and more precise than asynchronous counters because they don't have to wait for the signal to propagate through the flip-flops.

166 Part II Digital

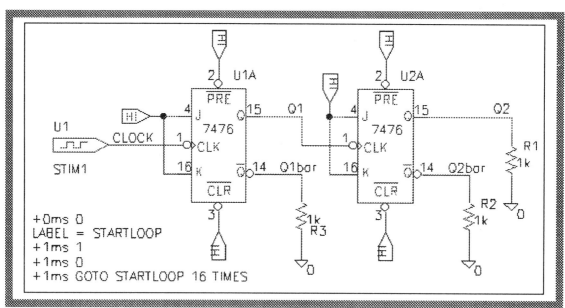

(a) MOD-4 binary ripple counter

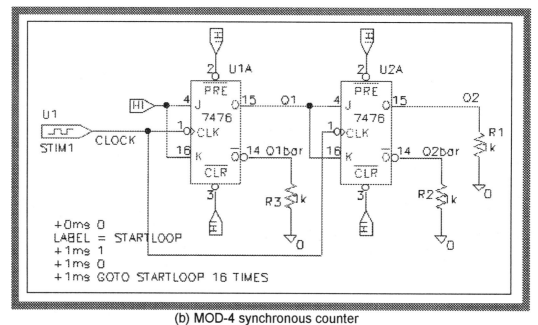

(b) MOD-4 synchronous counter

FIGURE 16.1

Two-stage binary counters

PSpice for Windows

Integrated circuit counters

Fortunately, a wide variety of integrated circuit counters are available to simplify circuit design. For 4-bit counters, the most popular modulus numbers are 10 and 16. The MOD 10 (BCD) counters use internal feedback to reset the count upon reaching 11.

In this chapter, after preliminary work with simple discrete MOD-4 counters, we will work primarily with the 7490 MOD-10 binary ripple counter, and the 74190 MOD 10 synchronous counter.

SIMULATION PRACTICE

The MOD-4 binary ripple counter

1. Draw the MOD-4 (2-bit) binary ripple counter circuit of Figure 16.1(a) and set the attributes as shown.

2. On the graph of Figure 16.2, predict (draw) the Q1, Q2, and Vout waveforms in the space provided. (Suggestion: Use pencil.)

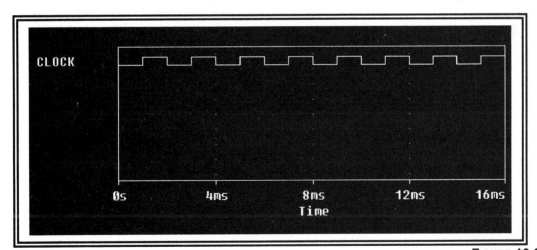

FIGURE 16.2

MOD-4 ripple counter waveform

3. Run PSpice and generate the CLOCK, Q1, and Q2 (Q2$DtoA) waveforms. Were they the same as your predictions? (Note: Be sure that the timing is *Typical* by: **Analysis, Setup, Digital Setup, Typical, OK, Close**.)

    Yes   No

4. Based on the waveforms of Figure 16.2, what is the frequency relationship between the clock input and the Q2 output?

    f(CLOCK) = _____ × f(Q2)

The MOD-4 synchronous counter

5. Repeat steps 1 through 4 for the synchronous counter of Figure 16.1(b). Were the results the same?

    Yes   No

The 7490A MOD-10 BCD ripple counter

6. Draw the asynchronous counter/decoder combination circuit of Figure 16.3 and set all attributes as shown.

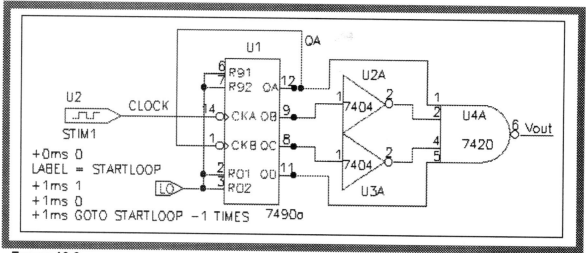

FIGURE 16.3

7490A ripple counter

7. Generate the graph of Figure 16.4. Draw a vertical line on the graph at the point where the count recycles from 9 back to 0.

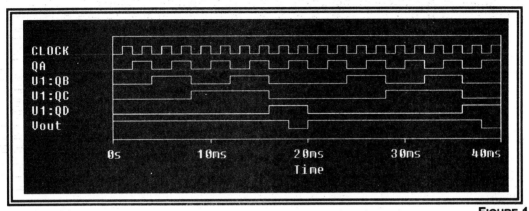

FIGURE 16.4
7490 waveforms

The 74160 MOD-10 BCD synchronous counter

8. Draw the synchronous counter circuit of Figure 16.5 and set all attributes as shown.

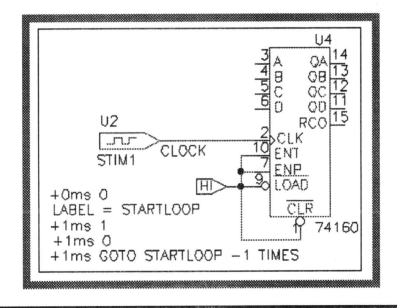

FIGURE 16.5
74160 synchronous counter

PSpice for Windows

9. Generate a set of output waveforms (QA, QB, QC, QD, and RCO).

   (a) Does the counter properly cycle from 0 to 9?

   Yes   **No**

   (b) Why is RCO called the "terminal count" pin?

10. Add a second cascaded 74160 to the circuit of Figure 16.5 to create the divide-by-100 synchronous counter of Figure 16.6.

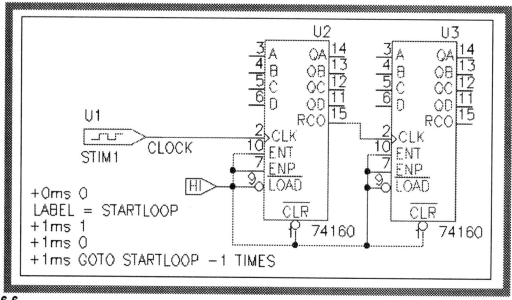

FIGURE 16.6

Cascaded synchronous counter

11. Generate a complete set of waveforms. Does U3:QD cycle once for every 100 CLOCK cycles?

   Yes   **No**

Advanced activities

12. Referring to the 7490A circuit of Figure 16.3, increase the clock frequency from 500Hz to 200MEGHz (change +1ms to +25ns). Switch to worst case analysis and generate the curves of Figure 16.7. (Caution: Be sure that *Print Step* is less than *Final Time*.)

    (a) Does the ambiguity increase from QA to QC as expected for a ripple counter?

    Yes   No

    (b) Carefully examining the 7490 internal circuitry in Appendix D, why doesn't the ambiguity increase from QC to QD?

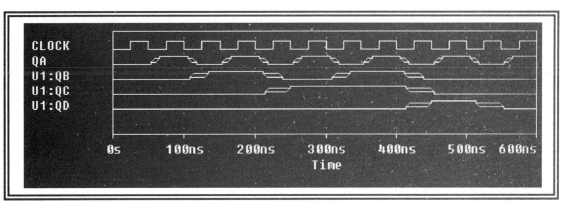

FIGURE 16.7
74160 synchronous counter waveforms

13. Increase the frequency further to 333MEGHz (change +25ns to +20ns). Based on the results of step 12 and this step, answer the following:

    Approximate maximum (hazard-free) clock frequency = _____

    Type of hazard when frequency > maximum = _____

14. Repeat steps 12 and 13 for the 74060 synchronous counter of Figure 16.5 and answer the following:

   (a) Why doesn't the ambiguity increase as we go from QA to QD?

   (b) What is the approximate maximum clock frequency?

   (c) What hazard occurs when the frequency > maximum?

EXERCISES

- Determine the range of the count (the beginning and ending counts) and the modulus of the asynchronous counter of Figure 16.8. (See Appendix D for the internal connections of the 7493 4-bit binary ripple counter.)

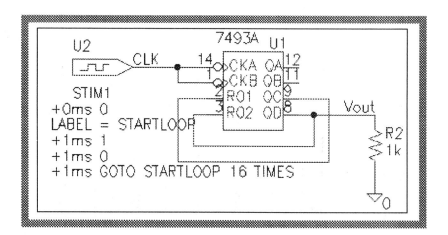

FIGURE 16.8

Reduced modulus ripple counter

- Determine the range of the count and the modulus of the synchronous counter of Figure 16.9. (See Appendix D for the internal connections of the 74160 4-bit synchronous counter.)

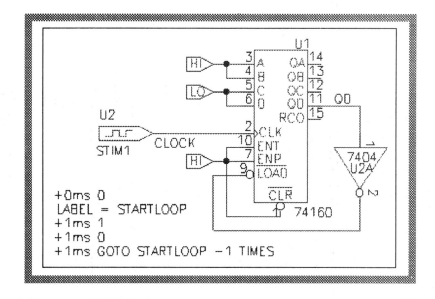

FIGURE 16.9
Reduced modulus synchronous counter

- Investigate the properties of the *Johnson counter* of Figure 16.10. What is the purpose of the RC circuit?

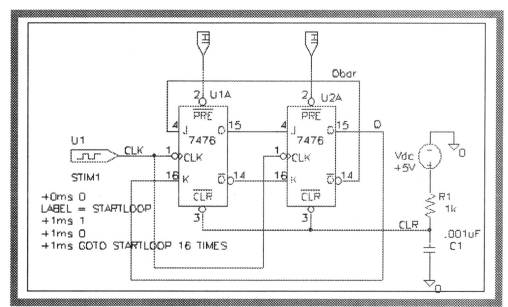

FIGURE 16.10
Johnson counter

- Draw the D-to-A converter of Figure 16.11, use clock speeds of 500Hz and 250MEGHz to generate the curves of Figure 16.12, and comment on the results.

- In Figure 16.11, change the 7490A asynchronous counter to the 74160 synchronous counter, drive the circuit at 250 and 333 MEGHz, compare the resulting waveforms to Figure 16.12, and comment on the results.

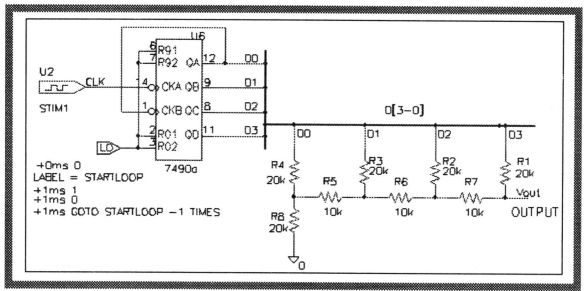

FIGURE 16.11

D-to-A converter

QUESTIONS & PROBLEMS

1. How many flip-flops would be required to design a counter that counts from 0 to 63?

2. Related to counters, what is the difference between *synchronous* and *asynchronous*?

Chapter 16 Counters   175

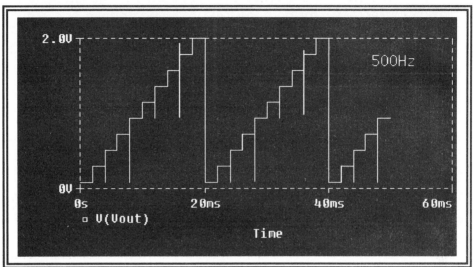

(a) 500Hz waveforms

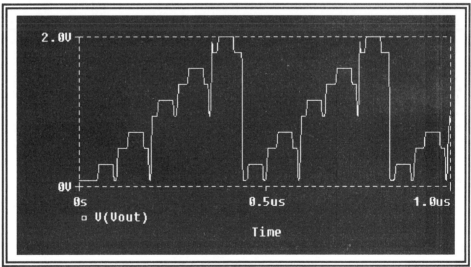

(b) 250MEGHz waveforms

FIGURE 16.12

A-to-D waveform comparison

PSpice for Windows

3. Because of the minimum WIDTH requirements of the CLOCK, the maximum frequency for both the 7490A asynchronous and 74060 synchronous counters were the same. What then is the advantage of the synchronous counter? (Hint: Review the second exercise.)

4. Based on Figure 16.9, design a MOD 7 (2-8) counter.

CHAPTER 17

Digital Coding
Comparators, Coders & Multiplexers

OBJECTIVES

- To design and test circuits that use comparators, encoders, decoders, and multiplexers.

DISCUSSION

In this chapter we investigate three popular types of digital integrated circuits:

- *Comparators*, which input two digital numbers and report if the first is greater than, equal to, or less than the second.

- *Encoders*, which convert a single line to a multiple-bit code, and *decoders*, which reverse the process.

- *Multiplexers*, which select one signal from a number of choices

SIMULATION PRACTICE

Comparators

The 7485 *4-bit magnitude comparator* of Figure 17.1 compares the values of the A and B inputs and activates the appropriate output line (A<B, A=B, A>B). When using a single 7485 (not cascaded), the A<B_IN, A>B_IN, and A=B_IN input signals are set as shown.

1. Draw the circuit of Figure 17.1 and set the attributes as shown.

PSpice for Windows

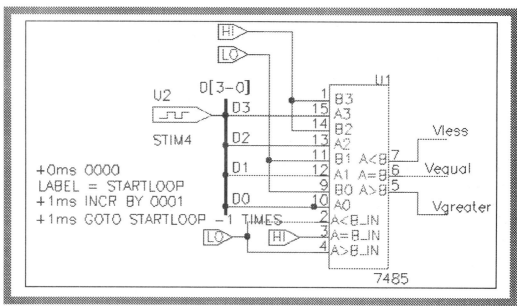

FIGURE 17.1

The 7485 comparator

2. Given the input waveforms of Figure 17.2, predict (draw) the output waveforms (*Vless*, *Vequal*, and *Vgreater*) in the space provided.

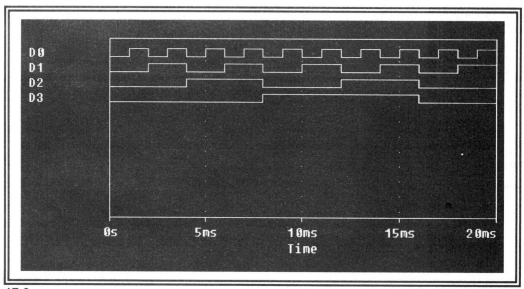

FIGURE 17.2

Comparator waveforms

PSpice for Windows

3. Generate the waveforms using PSpice and compare to your predictions. Are they the same? (Make any necessary corrections.)

    Yes   No

Encoders

The 74147 *10-line-to-4-line priority encoder* of Figure 17.3 converts decimal to BCD. It accepts any one of the nine active-low input lines representing decimal 1-9 and outputs the corresponding active-low BCD code. When all inputs are high, a decimal 0 is input. If more than one line goes active at a time, only the highest priority line—ranked from IN9 to IN1—is accepted.

4. Draw the encoder/DtoA circuit of Figure 17.3 and set the attributes as shown.

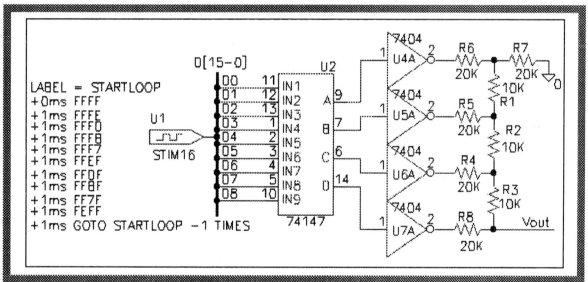

FIGURE 17.3

74147 Encoder test circuit

5. Given the input waveforms of Figure 17.4, predict (draw) the analog output waveform (Vout) in the space provided.

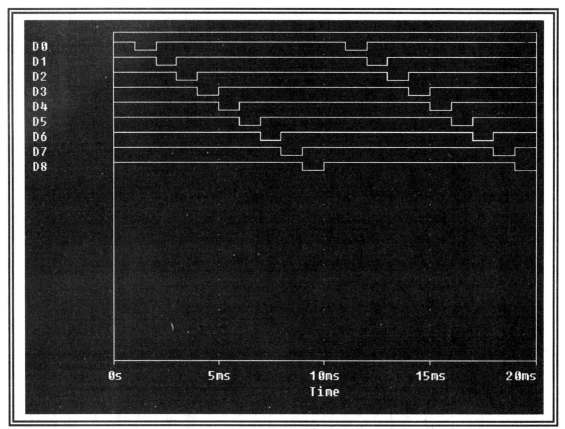

Figure 17.4

74147 Encoder waveforms

6. Generate the analog output waveform using PSpice and compare to your predictions. Are they approximately the same?

    Yes   No

Decoders

The 7442A *BCD-to-decimal decoder* of Figure 17.5 accepts four active-high BCD inputs and activates one of 10 active-low output lines.

7. Draw the decoder circuit of Figure 17.5 and set the attributes as shown.

8. Given the input waveforms of Figure 17.6, predict (draw) the output waveform (Vout) in the space provided.

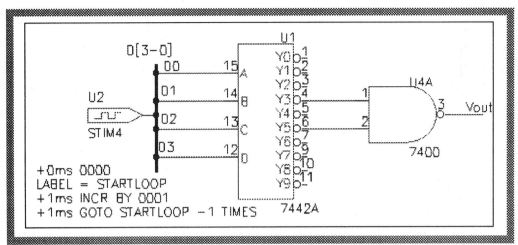

FIGURE 17.5

7442A decoder test circuit

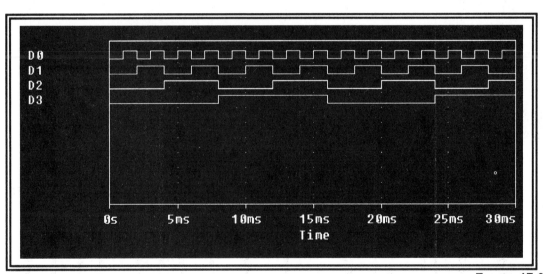

FIGURE 17.6

7442A decoder waveforms

9. Generate the waveforms using PSpice and compare to your predictions. Are they the same? (Make any necessary corrections.)

    Yes   **No**

Multiplexers

The 74151A *8-input multiplexer* of Figure 17.7 accepts a 3-bit code (from 000 to 111 on lines S0-S2) and routes the corresponding one-of-eight input lines (I0 to I7) to the single-line output (Z).

10. Draw the multiplexer circuit of Figure 17.7 and set the attributes as shown.

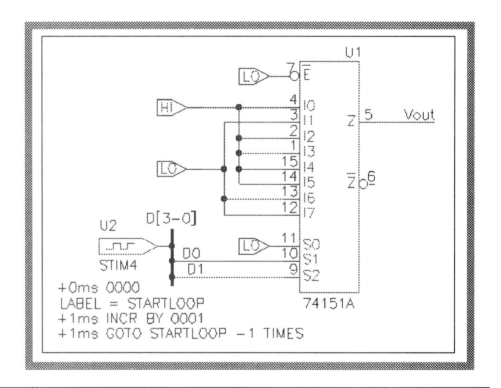

FIGURE 17.7

74151A test circuit

11. Given the input waveforms of Figure 17.8, predict (draw) the output waveform (Vout) in the space provided.

12. Generate the waveforms using PSpice and compare to your predictions. Are they the same? (Make any necessary corrections.)

    Yes   No

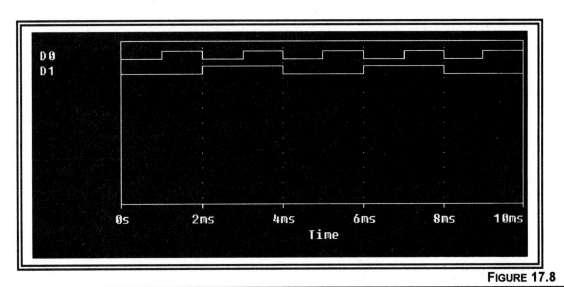

FIGURE 17.8
74151A test circuit waveforms

Advanced activities

13. What kind of combinational logic array does the multiplexer circuit of Figure 17.7 simulate? (Hint: consider S1 and S2 to be the inputs, and Z to be the output.)

EXERCISES

- Using worst case analysis, determine the highest "safe" frequency of operation of the decoder circuit of Figure 17.5.

QUESTIONS & PROBLEMS

1. What is the difference between an *encoder* and a *decoder*?

2. What type of device might to used to send the digital data from 10 sources down a single transmission line?

3. How would the output waveform change if the inverters of Figure 17.3 were bypassed?

4. Why is the 74147 called a *priority* encoder? (Hint: examine the 74147's data sheet in Appendix D.)

CHAPTER 18

The 555 Timer
Multivibrator Operation

OBJECTIVES

- To configure the 555 timer as an astable and monostable multivibrator.
- To configure the 555 timer as a voltage-controlled oscillator.

DISCUSSION

The 555 timer is another popular analog integrated circuit (like the op amp). This versatile chip is used for a wide variety of oscillation and timing needs. The internal schematics of the 555 are shown on Figure 18.1. Note that the 555 uses a combination of analog and digital parts.

By adding external components, the 555 can be configured as any of the following:

* Astable multivibrator
* Monostable multivibrator
* Voltage-controlled oscillator

SIMULATION PRACTICE

Astable multivibrator

1. Draw the circuit of Figure 18.2 and set the attributes as shown.

PSpice for Windows

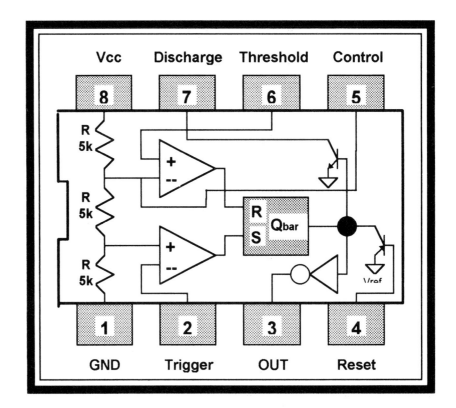

FIGURE 18.1

The 555 timer schematics

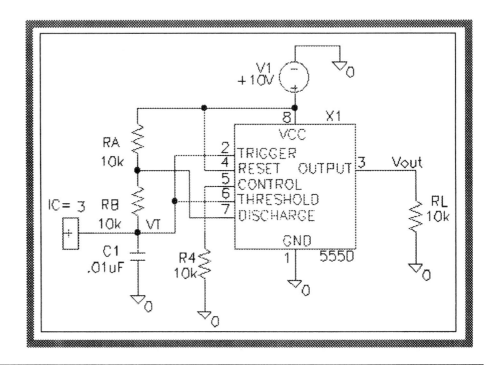

FIGURE 18.2

The astable multivibrator

PSpice for Windows

2. Using the equations below, calculate the expected frequency and duty cycle. Using the results, draw the predicted trigger (VT) and output (Vout) waveforms on the graph of Figure 18.3.

$$f_0 \text{ (frequency)} = \frac{1.44}{(R_A + 2R_B)C1} = \underline{\hspace{2cm}}$$

$$DtC \text{ (\% duty cycle)} = \frac{(R_A + R_B)}{(R_A + 2R_B)} \times 100 = \underline{\hspace{2cm}}$$

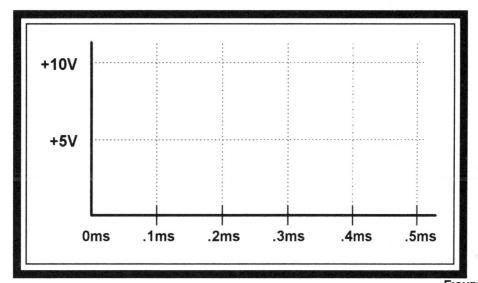

FIGURE 18.3
555 astable waveforms

3. Run a transient analysis (0 to .5ms) and draw the trigger and output waveforms on the graph of Figure 18.3. Do the actual waveforms approximately match the expected values? (Suggestion: Use IC1 to initialize C1 to approximately 3V, as shown.)

    Yes   No

Monostable multivibrator

4. By freeing up the trigger (pin 2), we configure the system for monostable operation (Figure 18.4).

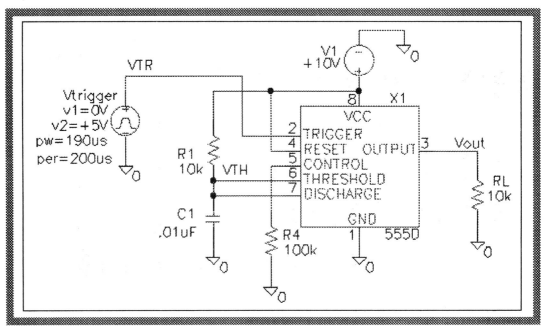

FIGURE 18.4

Monostable configuration

5. When the trigger goes low, the output goes high for a period of time determined by the equation PW = 1.1RC. Using this equation, draw the expected trigger (VTR), threshold (VTH), and output (Vout) waveforms on the graph of Figure 18.5.

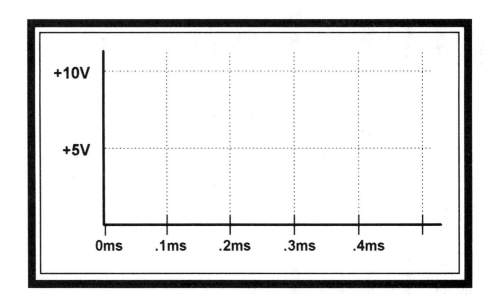

FIGURE 18.5

Monostable waveforms

PSpice for Windows

6. Run a transient analysis (0 to .5ms), and draw the actual trigger, threshold, and output waveforms on the graph of Figure 18.5. Be sure to properly label all waveforms. Do the actual waveforms approximately match the expected values?

    Yes   No

Advanced activities

7. By varying the voltage into pin 5, we create a voltage-controlled oscillator (VCO). Set up the 555 for VCO operation as shown in Figure 18.6.

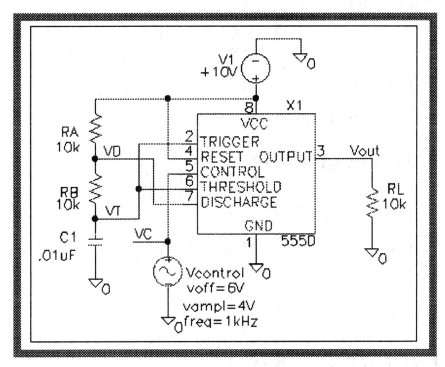

FIGURE 18.6
The 555 as a VCO

8. Run PSpice and generate the control (VC) and output (Vout) waveforms of Figure 18.7.

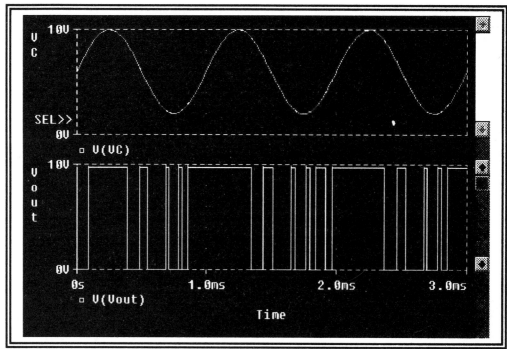

FIGURE 18.7
VCO control and output waveforms

9. Based on the results *approximately* what output frequency corresponds to each of the following control voltages?

(a) 2V

   f(out) @ 2V ≅ _____

(b) 6V

   f(out) @ 6V ≅ _____

(c) 10V

   f(out) @ 10V ≅ _____

EXERCISES

* Using the 555, design a circuit to delay exactly 1 second from the time of a trigger.

QUESTIONS & PROBLEMS

1. What does *monostable* mean?

2. Referring to Figure 18.1, what trigger and threshold voltages (as a percentage of VCC) will set and reset the flip-flop?

3. Is the trigger signal of the monostable multivibrator of Figure 18.4 active high or active low?

4. Which of the following signals reacts to a rising high-level voltage?
    (a) Threshold
    (b) Trigger

5. Referring to Figure 18.1, how does the control voltage (pin 5) of Figure 18.6 affect the frequency of operation?

PART III
Filter Synthesis

In most cases, throughout this text, we have first designed a circuit then tested its characteristics with PSpice.

However, with filter design, the process is usually reversed: we first specify the characteristics, then design the circuit to meet these specifications.

In this part we will use the special *Filter Synthesis* evaluation software (fseval54) that comes from MicroSim Corp. The evaluation version is identical to the professional version, except that all filters are limited to third order.

CHAPTER 19
Passive Filters
Biquads

OBJECTIVES

- To design a second-order Butterworth LC low-pass filter using the special filter synthesis evaluation package.

DISCUSSION

A filter is a circuit that selectively passes certain ranges of frequencies and attenuates others. There are four major types of filters:

- Low-pass — passes low frequencies, blocks high.
- High-pass — passes high frequencies, blocks low.
- Bandpass — passes a frequency range, blocks frequencies above or below the range.
- Band-stop (reject) — blocks a frequency range, passes frequencies above or below the range.

The exact input/output characteristics of each type of filter are given by a *transfer function*, which is a complex-number mathematical description of the output divided by the input. Highly accurate transfer functions require very complex mathematical formulations for their description. However, transfer functions can be successfully *approximated* by limiting the mathematical description to quotients of polynomials:

$$F(s) = \frac{a_0 + a_1 s + a_2 s^2 + \ldots + a_n s^n}{b_0 + b_1 s + b_2 s^2 + \ldots b_n s^n} \quad \text{(where } s = jw = j2\pi f\text{)}$$

PSpice for Windows

The order of the denominator (n) is called the "filter order." The higher the order, the more closely the approximation fits the ideal. Loosely speaking, each order corresponds to an inductor or a capacitor in the resulting circuit.

Because the polynomial form of the transfer function is difficult to work with, the polynomials are usually factored to yield even and odd order linear and quadratic terms:

Even order functions

$$F(s) = \prod_{i=1}^{n} \frac{k_2 s^2 + k_1 s + k_0}{s^2 + s\frac{w_0}{Q} + w_0^2}$$

Odd order functions

$$F(s) = \left(\frac{k_1 s + k_0}{s + w_0}\right) \left(\prod_{i=2}^{n} \frac{k_2 s^2 + k_1 s + k_0}{s^2 + s\frac{w_0}{Q} + w_0^2} \right)$$

Circuits that use such functions are called "biquads". The roots of the denominator (the complex frequencies that make the denominator zero) are termed the "poles" of the filter because the complex gain goes to infinity. The roots of the numerator are termed "zeros" because the complex gain goes to zero.

By selecting appropriate coefficients and terms in the biquad transfer function, we can design filter approximations that fall into five "classic" types:

- Butterworth
- Bessel
- Chebyshev
- Inverse Chebyshev
- Elliptic

PSpice for Windows

In this chapter, we will select the popular Butterworth ("maximally flat") type to generate frequency responses that are flat and smooth. To set a specific goal, we will design a *passive* LC filter that follows the ideal Bode characteristics of Figure 19.1. (In the following chapter, we will switch to an *active* filter.)

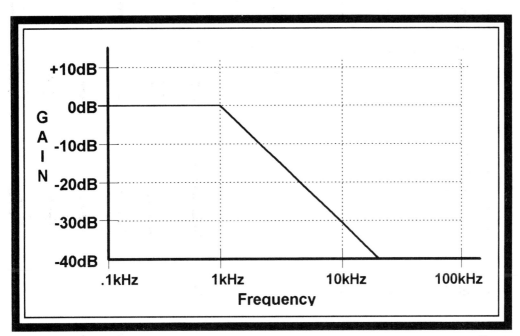

FIGURE 19.1
Desired frequency response

SIMULATION PRACTICE

Installation and execution

1. Install the Filter Synthesis Evaluation Package by following the directions on the diskette (run *install* from Windows).

2. Under the *File Manager* (Windows), run *filter.exe* in directory *fseval54* (**File**, **Run**, **OK**) and observe the main menu shown below:

    File Specify Coefficients Circuits Plots Preferences

Filter design

3. From the main menu, select *Preferences* (press "r" or use the arrow keys to highlight Preferences, <**Enter**>). Review *Filter Synthesis Note 19.1* to learn the meaning of the preferences options available.

Filter Synthesis Note 19.1
What is the meaning of the various preferences options?

- UNITS specifies the frequency multiplier for all frequency values.
- SAMPLED DATA specifies the clock frequency when performing switched capacitor filtering.
- S TO Z TRANSFORM specifies the method used when converting an s-transform function to a z-transform function.
- SPECIFY ALL STD. FILTERS BY gives us the option of specifying the filter characteristics by entering the stop band frequency or the order of the filter.
- SPECIFY BAND PASS/REJECT FILTERS BY gives us the option of specifying bandpass and band-reject filters by entering the center frequency and bandwidth, or the upper and lower band limits.
- SCF NETLIST FORMAT specifies the type of analysis to be performed for switched capacitor filter (SCF) circuits. PSPICE (normal) makes use of conventional circuit elements, PSPICE (ideal switch) uses switches, and SWITCAP uses switching capacitors.

---

4. Using the appropriate function keys, enter the following into the *Preferences* screen:

    Units: KHz
    S to Z transforms: mod bilinear
    Sampled data: Leave blank (erase if necessary)
    Specify all std filters by: stop band
    Specify band pass/reject filters by: center freq & bandwidth
    SCF netlist format PSPICE(normal)

    Press **F1** to save settings and return to the main menu.

5. Next, open the *Specify* menu, select *Low Pass*, <**Enter**>.

6. Following the requirements of Figure 19.1, enter the *Low Pass Limits*, and *Gain* as follows: (Use arrow keys to move the cursor.)

Pass band cutoff	1	KHz	(The 3 dB down frequency)
Stop band cutoff	10	KHz	(The frequency where the stop band attenuation is reached)
Pass band ripple	3	dB	(For a Butterworth filter, the ripple must be 3dB. The system will override any other value.)
Stop band atten.	30	dB	(The minimum attenuation in the output at the stop band cutoff frequency)
Gain	0	dB	(The gain at low frequencies)
Select	Normal Approx		

   Press F2 to select the Butterworth approximation. Note that the number of orders required for each of the filter types is automatically generated. As shown, our Butterworth design requires a second-order circuit.

7. Press ESC, ESC to return to the main menu and save all settings.

8. Next, enter the *Plot* menu, select the *Bode* option, and **<Enter>**. This plot shows the actual characteristics of the filter we have designed.

9. To compare our design to the goal of Figure 19.1, use the left/right arrow keys to move the cursor and verify the following: (Note the cursor position coordinates at the lower right of the screen.)

   (a) Is the gain approximately 0dB at low frequencies?

   Yes   No

   (b) Is the amplitude -3dB at the pass band cutoff frequency (1k)?

   Yes   No

   (c) Is the amplitude -30dB (*or better*) at the stop band cutoff frequency (10k)?

   Yes   No

10. Press **ESC** and return to the *Plots* menu.

11. Select *Pole-Zero*, **Enter**, and note the graph. The two "Xs show the location of the two poles (roots of the denominator of the biquad s-plane equation). The X-axis is the real component and the Y-axis is the imaginary component. (This filter has no zeros.)

12. Press ESC and again return to the *Plots* menu. Select *Time Domain*, **Enter**, and note the graph. The curve is the time-domain response of the filter circuit to a square wave input.

13. Using the cursor, measure the risetime (Tr)—the time between 10% and 90% of the steady-state level—and use the equation below to predict the breakpoint (Fb).

$$Fb = \frac{.35}{Tr}$$

Is the result (Fb) close to 1kHz?

    Yes   No

14. Return to the main menu (press **ESC** twice). Select *Circuits* and bring up the circuits menu.

15. Select *LC Ladders*, **Enter**, followed by *Singly Terminated*, **Enter**. The table on the screen is a description of the circuit in terms of branches, types, and roles (series or shunt).

16. To better interpret this table, bring up the circuit schematic (press F2). By switching between ESC and F2, note that the 1st branch is indeed a series inductor of 11.24mH and the 2nd branch is a shunt capacitor of 2.25uF. Also note the default 50 ohm singly terminating resistor (which can be changed from the bottom of the table).

17. If desired, repeat steps 14-16 and select *doubly terminated*. Compare the two cases (single and double). When done, return to the single case.

18. Return to the main menu (**ESC, ESC**).

Advanced activities

19. From the main menu, bring up the *Coefficients* menu, **Enter**.

PSpice for Windows

20. Select *s Biquads* and bring up the *S Domain* table. Using the information given, we generate the biquad transfer function for the single-stage second-order filter we designed earlier:

$$F(s) = \frac{39.57M}{s^2 + s\dfrac{6.29}{.707} + w_0^2}$$

21. Return to the *Coefficients* menu (**ESC**) and select *s Poles & Zeros*. Note that all the roots are infinity and are therefore poles.

22. Press F3 and note the location of the poles. Does this match the pole/zero plot of step 11?

23. Return to the *Coefficients* menu and select *s Polynomial*. Using the information given, we generate the polynomial form of the transfer function:

$$F(s) = \frac{39.7M}{39.7M + 8,900s + s^2}$$

24. Finished with our filter study, we exit the system (**ESC**, **ESC**, **File**, **Quit**.)

EXERCISES

- Using the "sketch" of Figure 19.2 to help select the proper specifications, design a bandpass filter using a third-order LC ladder.

QUESTIONS & PROBLEMS

1. What is the difference between a *passive* and an *active* filter?

2. What is a *transfer function*?

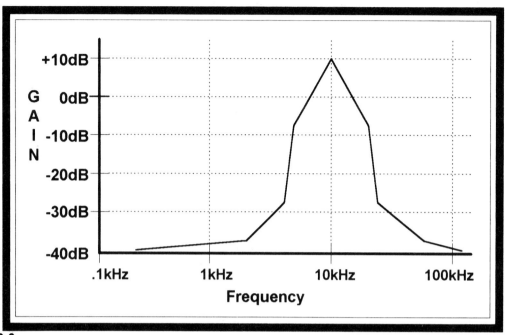

FIGURE 19.2

Bandpass design goal

3. What is the difference between the *polynomial* form and the *biquad* form of the transfer function?

4. Looking to Figure 19.1, we asked for a rolloff of 30dB/dec. Why then did the system generate a filter with a 40dB/dec rolloff? (Hint: what is the inherent rolloff of a second-order filter?)

5. What are *poles* and *zeros*?

PSpice for Windows

6. Based on this chapter, guess at the design of a third-order passive LC high-pass filter.

CHAPTER 20
Active Filters
The Switched-Capacitor Filter

OBJECTIVES

- To design a bandpass filter based on an active filter design.
- To introduce the *switched-capacitor* filter.

DISCUSSION

The *passive* filter of the last chapter contains only resistors, capacitors, and inductors. An *active* filter, on the other hand, is one that contains active elements—such as op amps. Compared to the passive filter, an active filter has "sharper" characteristics, can generate an overall gain, and does not require expensive inductors.

The switched-capacitor filter (SCF)

Another way to implement a biquad filter section is by way of switched capacitor circuits. To create such a filter, we start with a standard RC active biquad circuit and replace each resistor by a "charge pump" (a clocked switch and capacitor). Switched capacitor filters are the most common way to create a complete filter on a single chip.

The bandpass filter

The bandpass filter passes only a single frequency and blocks all others. In this chapter, we will design an active bandpass filter that passes only the tone-of-A (880Hz).

PSpice for Windows

SIMULATION PRACTICE

1. Enter filter designer main menu, use the left/right arrows to select the *Preferences* menu, press **Enter** to open the menu, and select the following:

   Units: Hz
   Sampled data: (Not applicable; leave blank; erase any numbers)
   S to Z transforms: mod bilinear
   Specify all std filters by: stop band
   Specify band pass/reject filters by: center freq & bandwidth
   SCF netlist format : PSPICE(normal)

   Press **F1** to save preferences and return to the main menu.

2. Open the *specify* menu, select *bandpass*, and enter the design parameters shown in Figure 20.1 in the appropriate sections (but do not enter **ESC** as yet).

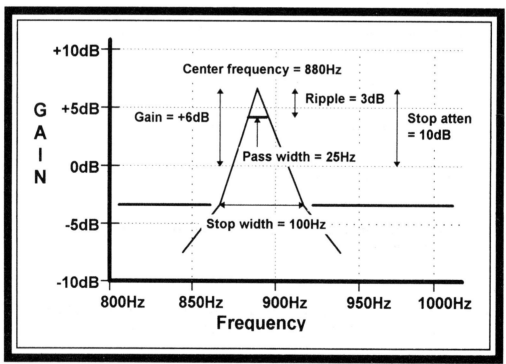

FIGURE 20.1
Desired frequency response

Chapter 20   Active Filters   207

3. Press **F2** to select the Butterworth approximation. Note that all filter types are second order.

4. Press **ESC** twice and return to the main menu.

5. Next, enter the *Plots* menu, select the *Bode* option, and press **Enter** to view the plot.

6. Use F2 to change the horizontal plot type to linear and press **Enter**.

7. To change the horizontal scale, use the up/down arrow keys to select *Min* and *Max*, change the *Min* and *Max* limits to 800 and 1000. Press **Enter** to command the system to accept the new limits.

8. Using the left/right arrow keys, move the cursor and verify that the plot meets the requirements of Figure 20.1:

   (a) Is the center frequency 880Hz?

       Yes   **No**

   (b) Is the gain 6dB (or better) at the center frequency?

       Yes   **No**

   (c) Is the pass band ripple 3dB?

       Yes   **No**

   (d) Is the pass band width 25Hz or less?

       Yes   **No**

   (e) Is the stop pass width 100Hz or less?

       Yes   **No**

   (f) Is the stop band atten 10dB (or better)? (<u>Remember</u>: We have a 6dB gain.)

       Yes   **No**

9. Return to the main menu (press **ESC** twice). Open the *Circuits* menu, select *RC Biquads*, *Basic MLF* (multiloop feedback), and display the schematic circuit (**F4**).

10. Looking at the circuit, are the resistors set to conventional values?

        Yes   **No**

PSpice for Windows

11. To set the resistors to conventional values, press **ESC** to return to the RC biquads screen. Press **F1** to begin the *round* process. Press **F2** to set the resistors to values commonly found in the 10% tolerance category; press **F3** to set the capacitors to values commonly found in the 5% tolerance category.

    Press **F1** to set, and notice the new (conventional) values. Press **F4** to select the schematic, and draw the circuit below:

12. By returning to the *RC Biquads* submenu each time, view the various circuit options listed below, and place an asterisk beside those that are *state* filters. (A *state* filter uses integrators and differentiators to solve the filter's time-domain differential equation.)

    Deliyannis-Friend
    Akerbery-Mossberg
    Tow-Thomas
    KUN

Switched Capacitor Filters

13. From the main menu, enter *preferences*. Set the clock frequency to 100kHz and *SCF netlist format* to SWITCAP. Press **F1** to set values and exit.

14. Enter the *Specify* menu, select *Band Pass*, and verify the specifications of Figure 20.1, select *Butterworth*, and return to the main menu.

15. Enter the plots menu, select Bode, and display the filter characteristics. Does it equal or better the specifications?

    Yes   **No**

16. Return to the main menu and enter the *circuits* menu. Select *Switched Cap Biquad* and press **F4** to view the circuit. Does the circuit contain only op amps and capacitors?

        Yes   **No**

17. Exit the filter synthesis program.

Advanced activities

18. Using PSpice or hands-on, test the new circuit design with any available operational amplifier (such as the UA741). Based on your test results, generate a Bode plot and compare to Figure 20.1.

19. Repeat the design using *Bessel, Chebyshev, or elliptic* types.

EXERCISES

- Design an active bandpass filter for a radio station of center frequency 1MEGHz and a pass width of 6K. To prevent overlap into neighboring stations, the stop width should be 12K at a stop attenuation of 10dB.

QUESTIONS & PROBLEMS

1. What are some of the advantages of an active filter design over a passive filter design?

2. Why is an operational amplifier used to implement analog filters? (Hint: does a filter make use of feedback?)

3. Why are switched capacitor filters normally implemented using commercial integrated circuits?

PART IV
Applications

In the five chapters of Part IV, we have an opportunity to put our skills to the test.

The five application projects consist of analog/digital combinational circuits that push the component limits of the evaluation version.

In all cases, we take full advantage of the modularization techniques introduced in Volume I.

Design Notes

Because the circuits designed within this applications section are larger than those in previous chapters, and because modularization techniques are used in all cases, we offer the following suggestions:

- Custom design alternative views for any or all of the modules (at outlined in Volume I, Chapter 29).

- To more quickly locate a node anywhere within the hierarchy for waveform display, simply drop a marker anywhere in the hierarchy and the corresponding trace shows up. Or, place the markers before simulation and perform **Analysis**, **Probe Setup**, **Show All Markers** (under *At Probe Setup*) for automatic display after simulation.

CHAPTER 21

Touch-Tone Decoding
The Bandpass Filter

OBJECTIVES

- To design, test, and modify a touch-tone decoder circuit.

DISCUSSION

Our first application is representative of those found in the modern-day *public switched telephone network* (PSTN). Individual subscriber phones connect to the *central office* (CO), which in turn routes calls to the proper destination.

Today's touch-tone telephones provide a keypad similar to that of Figure 21.1. When a key is pressed a *pair* of tones is generated. For example, when a "6" is pressed, tones 770Hz and 1477Hz are generated.

The two tones are summed and sent to the central office, where they pass through a series of filter-decoders to determine which key was pressed.

SIMULATION PRACTICE

1. Draw the top-level block design of Figure 21.2.

> The two frequencies of key 6 are summed by block SEND and transmitted to interconnect wire LOOP. Within block RECEIVE, the combined signal passes through filters corresponding to key 6. The outputs are ANDed together and presented at port SELECT.

PSpice for Windows

214 Part IV Applications

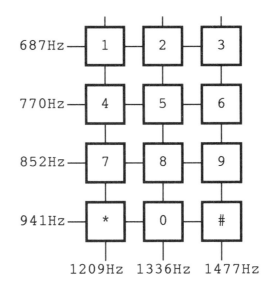

FIGURE 21.1
Touch-tone keypad

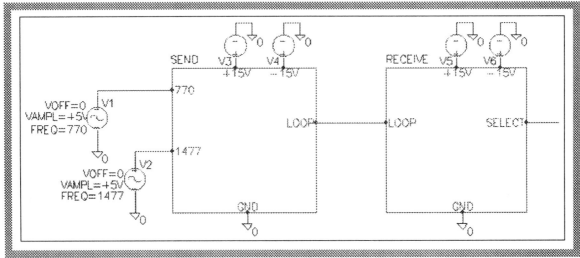

FIGURE 21.2
Touch-tone top-level design

PSpice for Windows

2. First, we PUSH into block SEND and create the design of Figure 21.3.

> An op amp summing circuit combines the two input signals and transmits them by way of line LOOP.

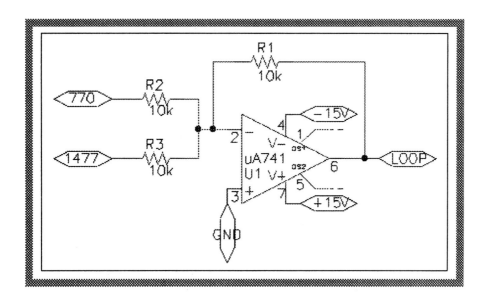

FIGURE 21.3
Circuit SUM

3. The SEND block done, we POP back to the top level and PUSH into block RECEIVE. This time, because the RECEIVE process is more complex, we create the mid-level block diagram of Figure 21.4.

> The received signal is filtered by block FILTER, the two outputs sent to block AND, and output from line SELECT.

4. We next PUSH into block FILTER and create the circuit of Figure 21.5.

> The summed (combined) signal from block SEND passes to bandpass filters U1 and U2 which are tuned to corresponding frequencies 770Hz and 1477Hz. The outputs (F770 and F1477) pass to block AND.

216 Part IV Applications

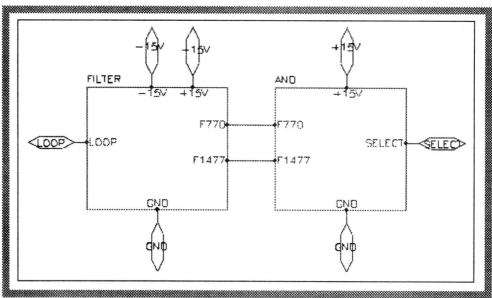

FIGURE 21.4

SEND mid-level block diagram

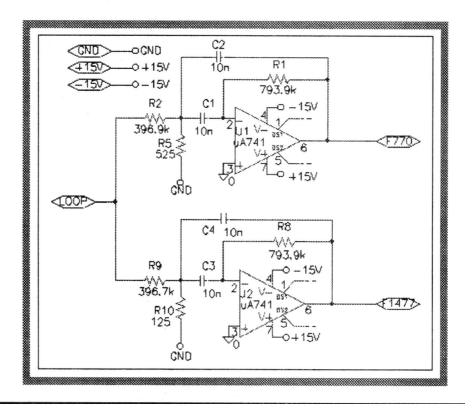

FIGURE 21.5

Block FILTER low-level design

PSpice for Windows

5. Finally, we POP back to the mid-level, PUSH into block AND, and create the circuit of Figure 21.6.

> The two filtered signals enter the diode AND gate and pass to the sample-and-hold circuit where the output is presented at port SELECT. If the combined transmitted signal matches the filters, the AND gate output is high and we have a "hit." Otherwise, the output is low and we have a "miss." (Note: we use a diode AND gate because of the limitations of the evaluation version.)

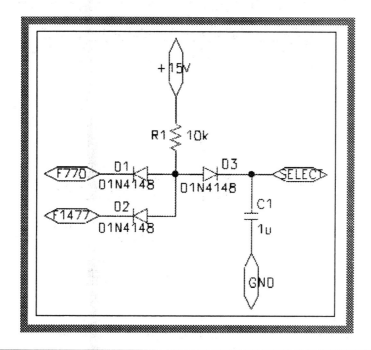

FIGURE 21.6
AND gate and sample-and-hold

6. Perform a transient analysis and display the output waveform of the SEND block in both time and frequency (Figure 21.7).

> *Note:* For smoother waveforms, be sure to set *Step Ceiling* to approximately period/50, or about 15μs in this case [1/(1477×50)].

Does the time-domain signal appear to be a composite of two frequencies?

    Yes   **No**

Do the frequency components of the combined signal peak at approximately 770Hz and 1477Hz, and are they of nearly equal amplitude?

    Yes   **No**

218 Part IV Applications

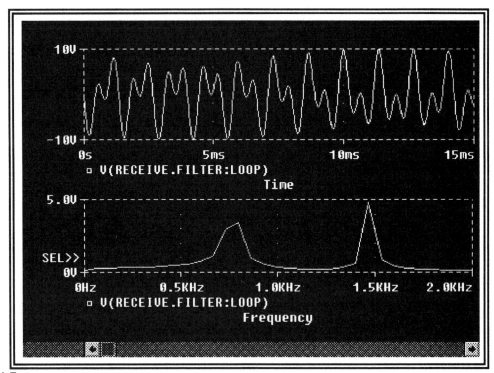

FIGURE 21.7
RECEIVE block waveforms

7. Next, display the output waveform of the RECEIVE block (output line SELECT), as well as the outputs of the two filters (F770 and F1477). (See Figure 21.8.)

   (a) Approximately how much time is required for the circuits to reach steady state?

   Time to reach steady state = _____

   (b) Are the two filter outputs high and of nearly equal amplitude?

         Yes   **No**

   (c) Is the output a "hit" (greater than 2V at steady state)?

         Yes   **No**

8. Change the input frequencies to match the key of 3 (change 770Hz to 687Hz), and again display the output waveforms (Figure 21.9).

   (a) Is the F770 output less than the F1477 output?

         Yes   **No**

   (b) Is the output a "miss" (less than 2V)?

         Yes   **No**

PSpice for Windows

Chapter 21 Touch-Tone Decoding 219

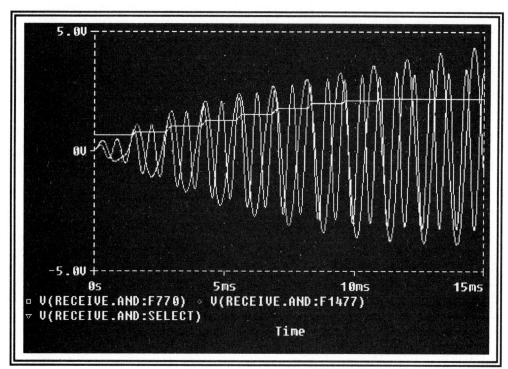

FIGURE 21.8

Key 6 output waveforms (HIT)

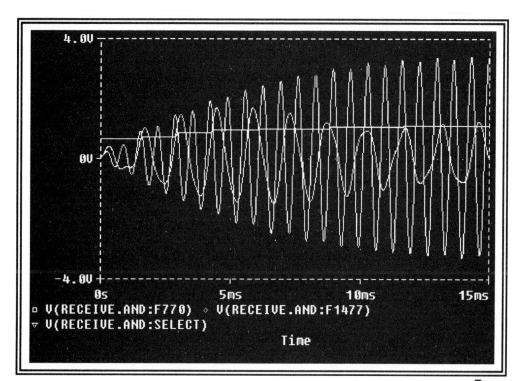

FIGURE 21.9

Key 3 output waveforms (MISS)

PSpice for Windows

Test and Modification Suggestions

Note: Due to the limitations of the evaluation version, most of the modifications in this and future applications chapters will have to be tested separately or evaluated "on paper." Report all findings on a separate sheet.

9. With the help of Figure 21.10, modify the output *sample-and-hold* circuit to include the ability to clear the output when the key is released.

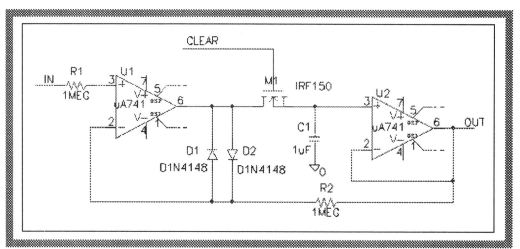

FIGURE 21.10

Sample-and-hold with clear

QUESTIONS & PROBLEMS

1. Which tones are generated when a "0" is pressed?

2. What advantages do active filters have over passive filters?

PSpice for Windows

3. Were it not for the limitations of the evaluation version, could we have used a conventional digital AND gate in the circuit of Figure 21.6, rather than the diode-based AND gate?

4. What function is performed by output components D3 and C1 of Figure 21.6?

5. Why is the bandwidth of the two filters an important factor in the design?

6. Explain how you might use the AC Sweep mode to test the filter characteristics of the two bandpass filters.

CHAPTER 22
Pulse-Code Modulation
Time-Division Multiplexing

OBJECTIVES

- To design, test, and modify a pulse code modulation (PCM) circuit.

DISCUSSION

In the modern-day PSTN (public switched telephone network), individual phones connect to *central offices* (COs). These COs are in turn interconnected by way of *trunk* lines, which normally consist of many individual copper wires bundled together.

For efficiency, there is a need to have each individual physical wire carry more than one telephone call. This is accomplished by a process called *multiplexing*, the subject of this chapter.

> *Multiplexing* is the process of combining two or more signals onto a single transmission line.

FDM versus TDM

There are two major types of multiplexing:

- FDM (frequency-division multiplexing), in which two or more signals are placed in different *frequency* slots.

- TDM (time-division multiplexing), in which two or more signals are placed into different *time* slots.

PSpice for Windows

224   Part IV   Applications

At the present time, the older FDM system is rapidly being replaced by the newer TDM system. When TDM includes conversion of each time-sampled analog signal to a serial digital stream, we have a PCM (pulse-code modulation) system—the subject of this chapter.

In our simplified PCM system, two input analog voice channels are continuously sampled in sequence. Each analog signal is converted to a 4-bit digital word and transmitted in serial by time-multiplexing.

At the destination, the process is reversed and the serial digital words are demultiplexed, converted back to analog, and output on two analog lines.

SIMULATION PRACTICE

1. Draw the top-level block design of Figure 22.1.

> Inputs TALK1 and TALK2 represent two phones simultaneously transmitting analog signals. The two analog inputs are converted to a PCM digital stream by block TRANSMIT and transmitted by way of line TRUNK. The digital data stream is converted by block RECEIVE back to the original analog signals and output to lines LISTEN1 and LISTEN2. Digital stimulus STIM1 provides the basic clock signal for timing all operations.

2. First, we PUSH into the TRANSMIT block and create the mid-level block design of Figure 22.2.

> Block PAM (Pulse-Amplitude-Modulation) samples the analog input signals, block PCM (Pulse-Code-Modulation) converts each analog signal to digital, and block P2S (Parallel-to-Serial) converts each 4-bit digital word to serial and transmits it to output line TRUNK.

PSpice for Windows

Chapter 22 Pulse Code Modulation

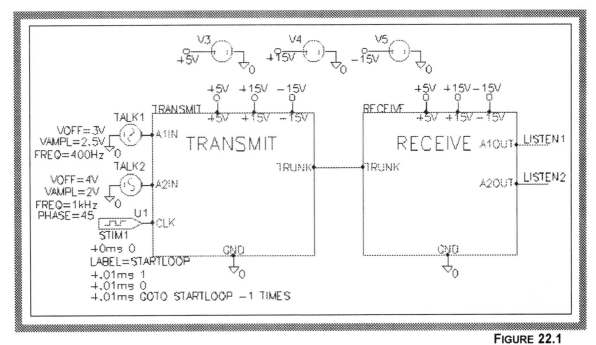

FIGURE 22.1
PCM top-level design

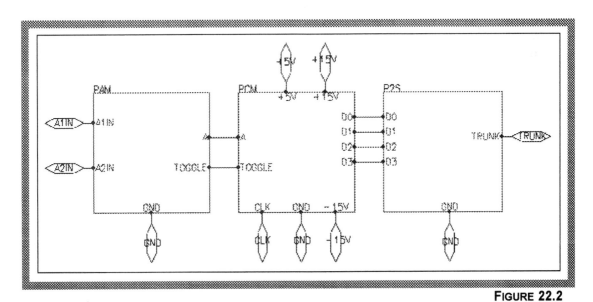

FIGURE 22.2
PCM TRANSMIT mid-level design

3. From mid-level block TRANSMIT, we first push into the PAM block and create the bottom-level design of Figure 22.3.

> The two analog inputs (A1IN and A2IN) are sampled (multiplexed) and presented at the analog output (A). Input TOGGLE times the multiplex rate by successively clocking the flip-flop and shorting the input signals through M1 and M2.

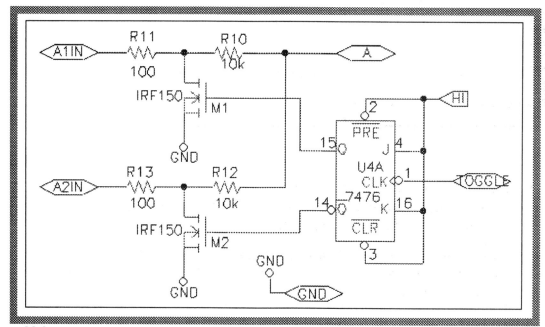

FIGURE 22.3
PAM low-level circuit

4. Next, we return to the mid-level, push into the PCM block and create the circuit of Figure 22.4. (For display convenience, be sure to tag the appropriate lines DAC and LATCH.)

> The analog input (A) is converted to digital by the counter technique, and presented at the digital outputs (D0, D1, D2, and D3) by the LATCH signal from the comparator. Each conversion process generates a TOGGLE signal, and the process continues. Digital input CLK clocks the A-to-D process.

Chapter 22 Pulse Code Modulation 227

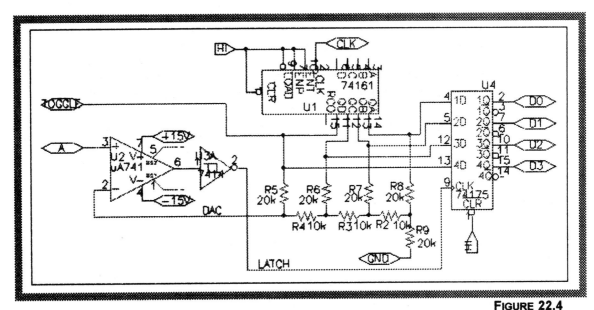

FIGURE 22.4

PCM low-level circuit

5. Again we return to the mid-level and push into block P2S. Unfortunately, the limitations of the evaluation version require us to leave this and all RECEIVE blocks empty.

> *Note*: Be sure to POP to the top, PUSH into block RECEIVE, and ground interface pins A1OUT and A2OUT to avoid floating (unbounded) pins.

6. Analyze the circuit and first show the input analog waveforms (A1IN and A2IN) of Figure 22.5. (We ignore the digital simulation errors for now.)

> If necessary, initialize all flip-flops to 0 by **Analysis**, **Setup**, **Digital Setup**, **All 0**, **OK**. (Timing should already be **Typical**.)

7. Next generate the internal waveforms of Figure 22.6.

> To expose the full digital names, drag the left-hand border of the digital display to the right. Or, go to the PROBE section of *msim.ini* in the Windows directory and add: DGTLNAMELEFTJUSTIFY=ON

   (a) Is the PAM circuit working properly?

       Yes   No

   (b) Is the PCM circuit working properly?

       Yes   No

PSpice for Windows

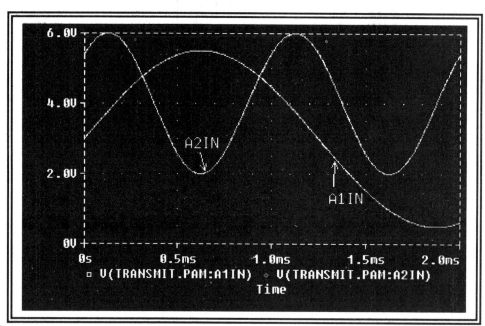

FIGURE 22.5

Input analog waveforms

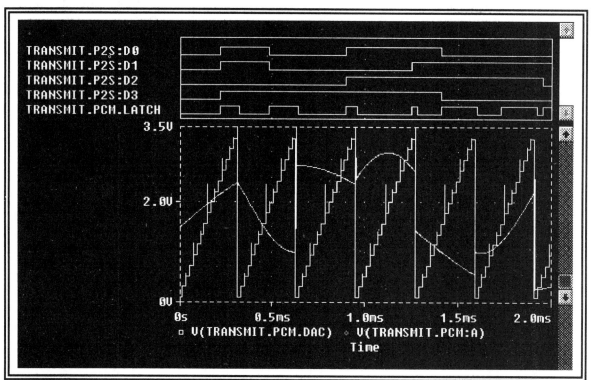

FIGURE 22.6

Internal PCM waveforms

PSpice for Windows

8. Does the system obey the *Nyquist* theorem? (Is the sampling rate for both A1IN and A2IN at least twice the input frequency?)

   Yes   No

Test and Modification Suggestions

Reminder: The present circuit has already reached the maximum component count for the evaluation version, and therefore most of these modifications will have to be made separately or "on paper." Report all findings on a separate sheet.

9. Complete the P2S block and convert each digital output signal to a serial data stream.

10. Complete the S2P, DEMUX, and FILTER blocks within the RECEIVE block. (See Figure 22.7 for a possible solution to the S2P and DEMUX blocks.)

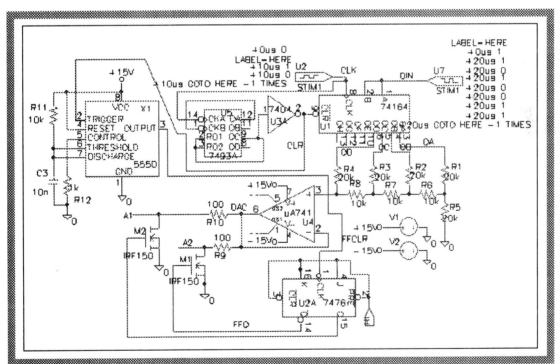

FIGURE 22.7
RECEIVE block circuits

11. View the digital hazards found in step 6 by **Tools, Simulation Messages**. The digital hazards that occur result from out-of-range voltages from U2 applied to U3 (Figure 22.4). How would you "clip" U2's output?

12. Assign the eight resistors of the D-to-A converter 10% tolerances and perform a Monte Carlo or worst case analysis.

QUESTIONS & PROBLEMS

1. List all the components that make up the D-to-A (digital-to-analog) converter.

2. Why are we guaranteed that switches M1 and M2 can never be ON at the same time?

3. According to the Nyquist rule, to avoid losing data, we must sample the input analog signals at twice the highest frequency of interest. Based on this, what is the highest analog input frequency that the circuit of Figure 22.3 can sample without losing data?

4. For high-speed operation, why is the 74161 synchronous counter preferable to the 7493A ripple counter?

5. For proper PCM operation, it is necessary that the analog signal sampling rate be constant. Was this achieved with the **TRANSMIT** circuitry? How was it achieved?

CHAPTER 23

Serial Communication
RS-232C

OBJECTIVES
- To design and test an RS232C serial communication system.

DISCUSSION

Our third application performs a classic duty: Computer-to-computer serial communication using the asynchronous RS-232 standard of Figure 23.1.

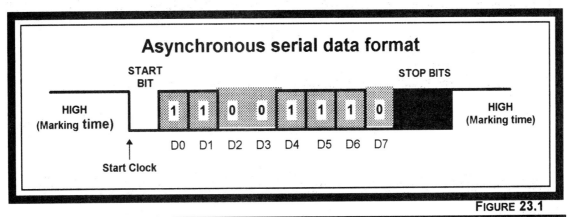

FIGURE 23.1
RS-232 asynchronous serial format

In order to synchronize the transmitter and receiver, each 8 bits of data are preceded by an active-low start bit and two active-high stop bits. When the receiver detects each high-to-low transition of the start bit, all clocks are reset.

PSpice for Windows

234 Part IV Applications

At the transmitting end, each 8 bits of parallel data are converted to serial, the start and stop bits factored in, and the serial data transmitted. At the receiving end, the process is reversed. The serial data is received, the start and stop bits stripped off, and the resulting serial data converted to parallel.

SIMULATION PRACTICE

1. Draw the top-level block design of Figure 23.2.

> 8-bit parallel data enters block TRANSMIT by way of stimulus input PIN (parallel-in). START and STOP bits are added and the resulting serial data transmitted from pin SERIAL. The RECEIVE block inputs the serial data, strips off the START and STOP bits, and outputs the parallel data from D0-D7. Input CLOCK times the overall system.

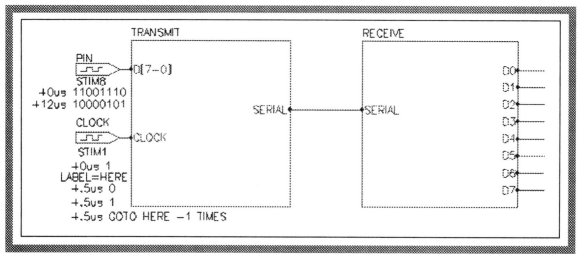

FIGURE 23.2
RS-232 top-level design

2. We PUSH into block TRANSMIT and create the mid-level design of Figure 23.3.

> Block COUNTER counts from 0 to 11. Block MUX accepts counter pulses and parallel input data and outputs serial data on several lines. Block GATES inputs the serial data and converts it to a single line.

PSpice for Windows

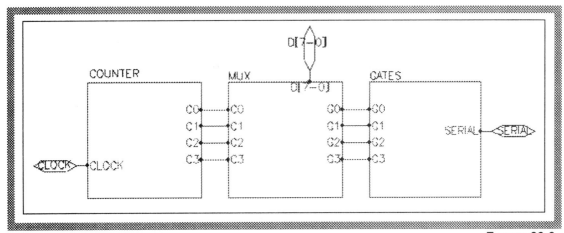

FIGURE 23.3
TRANSMIT
mid-level design

3. We PUSH into the COUNTER block and create the design of Figure 23.4.

> Block COUNTER uses a synchronous counter chip (74161) and feedback to count from 0 to 11 and output the count from lines C0-C3.

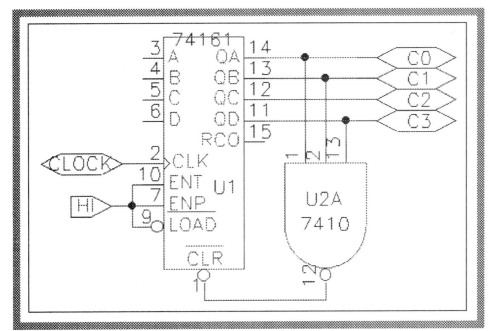

FIGURE 23.4
Block COUNTER design

PSpice for Windows

4. The COUNTER block done, we POP back to the mid level, PUSH into block MUX, and create the design of Figure 23.5.

> The central feature of this block is the two 8-in/1-out multiplexers. Clocked by the 0-11 inputs (C0-C3) from block COUNTER, the two 8-bit multiplexers are scanned. Starting at the top, we first output the START bit, then the 8 data bits, and finally the two STOP bits. The serial outputs from each multiplexer, as well as two enabling signals, are output to the next block.

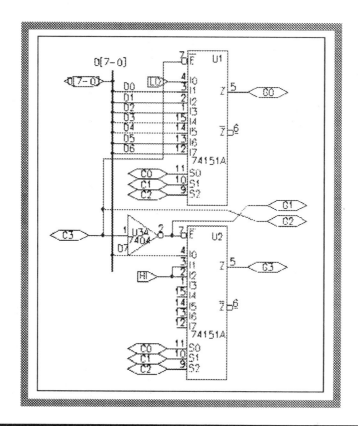

FIGURE 23.5

Block MUX design

5. Finally, we repeat the process to create block GATES of Figure 23.6.

> Block GATES ORs together the serial outputs from the two multiplexers. Steering logic G1 and G2 switches the serial flow as we switch from the upper multiplexer to the lower.

PSpice for Windows

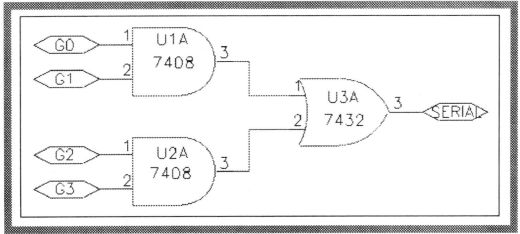

FIGURE 23.6

Block GATES design

6. Unfortunately, we cannot complete block RECEIVE due to the limitations of the evaluation version. However, we can't leave outputs D0-D7 with no connections. So we PUSH into block RECEIVE and short all the lines together.

7. Our design complete, we perform a transient analysis from 0 to 24µs and display the clock and serial output of Figure 23.7.

    (a) Using arrows or circles (or a method of your choice), identify the START bit and two STOP bits for each of the transmitted words.

    (b) Next, identify data bits 01110011 and 10100001 (D0-D7) for each of the two transmitted words.

    (c) Was D0 transmitted first, or was D7 transmitted first? (Circle your answer.)

        D0   D7

8. View additional signals of your choice. (For example, the counter output to check for 0-11 operation.)

PSpice for Windows

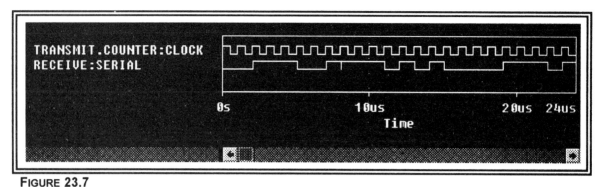

FIGURE 23.7
Block TRANSMIT clock and output waveforms

Test and modification suggestions

9. Perform a worst case digital analysis of the TRANSMIT circuit and generate the output shown in Figure 23.8. (**Analysis**, **Setup**, **Digital Setup**.)

   (a) What caused the hazard?

   (b) What action can you take to eliminate the hazard?

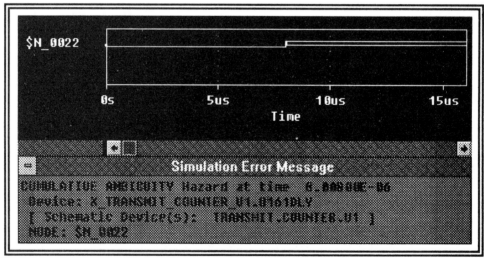

FIGURE 23.8
Block TRANSMIT worst case hazard

10. Modify the TRANSMIT circuit to include only a single STOP bit.

11. Modify the TRANSMIT circuit so that if no parallel input is presently available, a HIGH (MARK) is continuously transmitted until the next parallel data byte is available.

12. Design a circuit to carry out the RECEIVE block. (*Hint*: Modularize the circuit of Figure 23.9.)

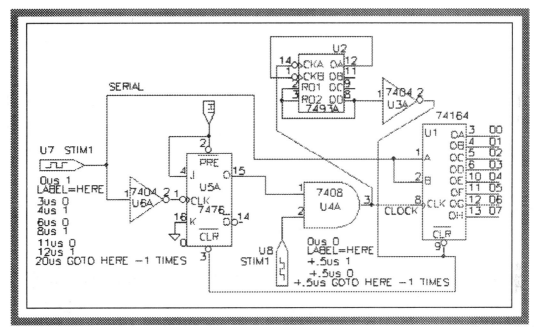

FIGURE 23.9
Suggested RECEIVE design

13. Modify the RECEIVE circuit so the parallel output data is latched each time it is available.

14. Redesign the TRANSMIT circuit of Figure 23.3 using a parallel-to-serial chip (not available in the evaluation version).

15. Redesign the RECEIVE circuit so the start bit resynchronizes the clock. (<u>Hint</u>: Use a 555 timer.)

PSpice for Windows

QUESTIONS & PROBLEMS

1. During RS-232 asynchronous communication, what is the purpose of the START bit? Of the two STOP bits?

2. What is the major advantage of serial communication over parallel communication?

3. As each serial byte is transmitted, what is the percentage of "overhead" (START and STOP bits) required?

CHAPTER 24

Big-Ben Chime Control
The Comparator

OBJECTIVES

- To design and test a digital circuit that controls clock chimes.

DISCUSSION

Many tower clocks (such as London's Big Ben), sound the time each hour by generating the corresponding number of chimes. A main clock continuously counts in hours from 1 to 12. At each hour transition, the corresponding number of chime pulses is sounded.

SIMULATION PRACTICE

1. Draw the top-level block design of Figure 24.1.

> Pulse trains HOUR and SECOND enter block BIGBENTOP and the output to the chimes appears at line BELL. Every hour, a series of chime signals equal to the hour (from 1 to 12) appears from BELL at the rate of 1/second.
>
> In order to shorten computation times during development, all clock sources have been greatly speeded up. For example, the 1 hour counter clocks at 2µs, and the 1 second counter clocks at .1µs (100ns).

PSpice for Windows

242 Part IV Applications

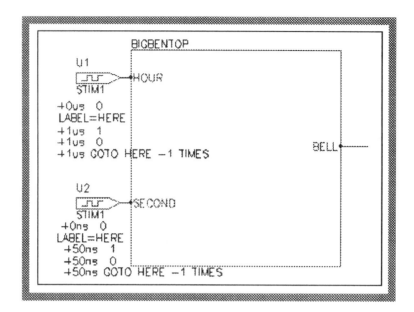

FIGURE 24.1

Big-Ben top-level design

2. We PUSH into block BIGBENTOP and create the mid-level design of Figure 24.2.

> **BLOCK HOURCTR** counts up at the rate of once per hour from 1 to 12 and outputs the time as a digital word on H0-H3. The hour clock is passed on to block SECONDCTR via line START.
> **BLOCK COMPARE** takes digital inputs from HOURCTR and SECONDCTR and outputs a CLEAR pulse when the two are equal.
> **BLOCK SECONDCTR** counts up from 1 to the specified hour at the rate of 1/sec when gated by START. The count value is sent to block COMPARE and the gate disabled when the count reaches the proper value and issues a CLEAR signal.

3. We PUSH into the HOURCTR block and create the design of Figure 24.3.

> Block HOURCTR uses a synchronous counter chip (74161) and feedback to count from 1 to 12 and output the count from lines H0-H3. Output START is sent to block SECONDCTR to begin the chime process.

PSpice for Windows

Chapter 24 Big-Ben 243

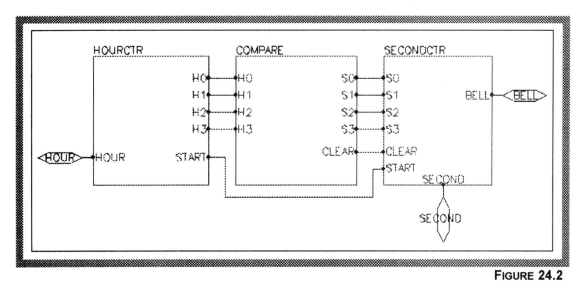

FIGURE 24.2
BIGBENTOP mid-level design

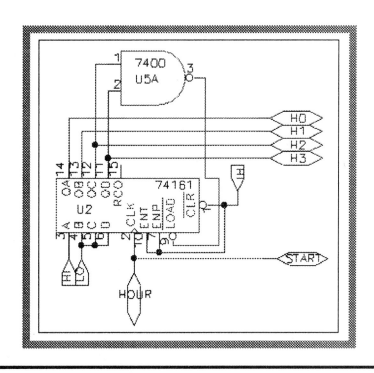

FIGURE 24.3
Block HOURCTR design

4. The HOURCTR block done, we POP back to the mid level, PUSH into block COMPARE, and create the design of Figure 24.4.

> Digital inputs H0-H3 (hours) count up by one each hour. At the change of each hour the seconds counter counts up at the rate of 1/sec. When the two counts are equal, a CLEAR pulse is sent to block SECONDCTR and the seconds count stops until the next hour transition.

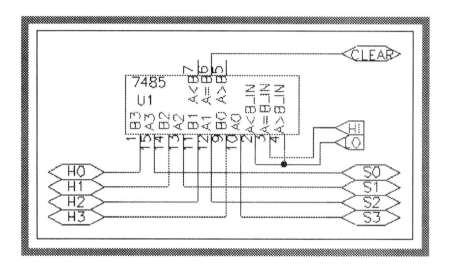

FIGURE 24.4
Block COMPARE design

5. Finally, we repeat the process to create block SECONDCTR of Figure 24.5.

> Block SECONDCTR receives the START signal, which sets the flip-flop and gates the second count out to BELL. Simultaneously, the 7493A counter counts up and sends the count to block COMPARE. When the second count equals the hour count, CLEAR is issued and the second pulses are blocked.

6. Our design complete, we perform a transient analysis from 0 to 30µs and display the output (BELL) curves of Figure 24.6. (The bottom curve is an "unsynched" blowup of hours 2, 3, and 4.)

Is the circuit functioning as expected?

Yes   No

Chapter 24 Big-Ben

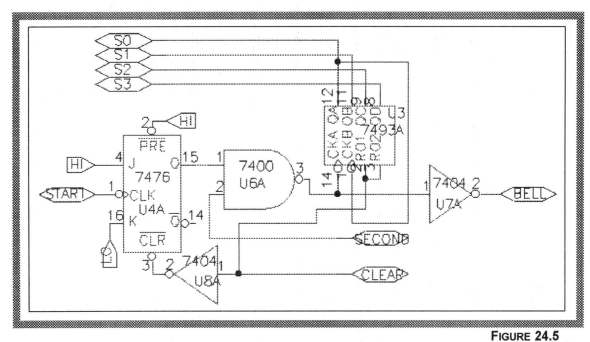

FIGURE 24.5
SECONDCTR design

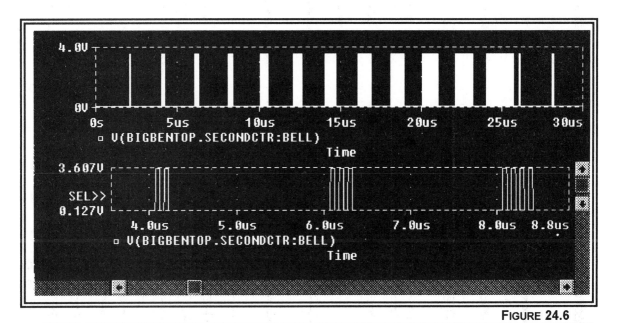

FIGURE 24.6
BELL output waveforms

Test and modification suggestions

7. Modify the design so only a "second" clock source is required.

8. Perform a worst-case analysis of the present circuit. Why would the cumulative ambiguity hazards uncovered be no problem when the circuit is placed into actual operation?

9. Modify the circuit to include an AM/PM indicator.

10. Following up on step 9, further modify the circuit so the chimes are inhibited during the 1AM to 5AM period (when most people are sleeping).

QUESTIONS & PROBLEMS

1. Although the hour counter counts up very slowly (in "real life"), why nevertheless did we choose a high-speed synchronous 74161 counter?

2. When is the CLEAR signal issued?

3. What is accomplished by the R01 and R02 inputs to the 7493A of Figure 24.5?

CHAPTER 25

Alarm Entry
The Encoder

OBJECTIVES

- To design and test a digital circuit that sets or disarms an alarm system upon entry of the proper sequential code.

DISCUSSION

To set or disable an alarm system, a popular method is to enter a coded sequence of numbers. If the sequence matches the code programmed into the system, a "set" or "open" signal is issued.

SIMULATION PRACTICE

1. Draw the top-level block design of Figure 25.1.

   > If the proper active-low sequence of code numbers is entered into "keys" ZERO, ONE, and TWO, OPEN goes high.

2. We PUSH into block ALARMTOP and create the mid-level design of Figure 25.2.

   > **BLOCK COMPARE** accepts entry and coded numbers and issues a RESET when they are not the same. CLOCK occurs with each entry.
   > **BLOCK CODE** accepts binary count inputs from C0 and C1, and in turn sends the coded entry sequence to block COMPARE.
   > **BLOCK COUNT** increments C0/C1 for every CLOCK cycle. The count goes back to zero upon a RESET signal. When the count reaches three (no RESET received), OPEN goes high.

PSpice for Windows

248 Part IV Applications

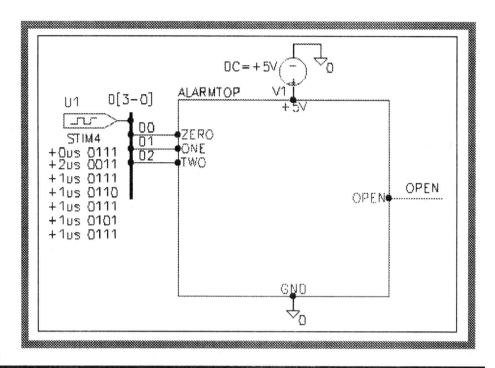

FIGURE 25.1
Alarm entry
top-level design

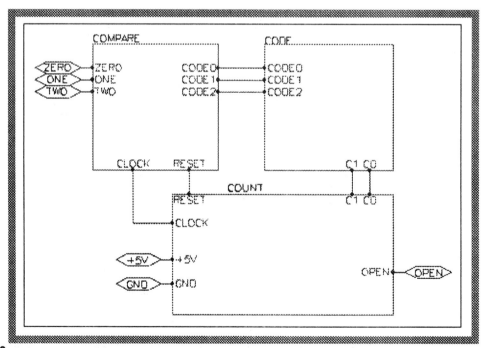

FIGURE 25.2
Block ALARMTOP
mid-level design

PSpice for Windows

3. We PUSH into block COMPARE and create the design of Figure 25.3.

> **BLOCK COMPARE** accepts the entry code and programmed code sequences and compares them. Whenever they do not compare (equal each other), a RESET is issued.

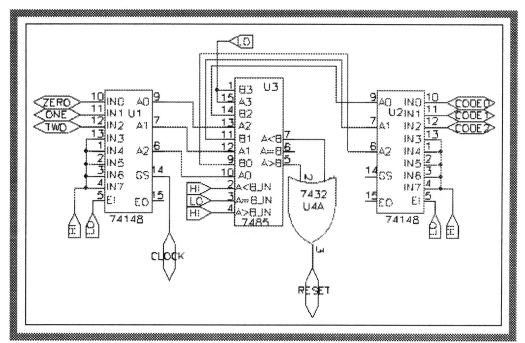

FIGURE 25.3
Block COMPARE design

4. We POP back to the mid level, PUSH into block CODE, and create the design of Figure 25.4.

> **BLOCK CODE** inputs a binary count from 0 to 3 from C0 and C1, and sequences CODE0, CODE1, and CODE2 according to the programming provided by the gate array.

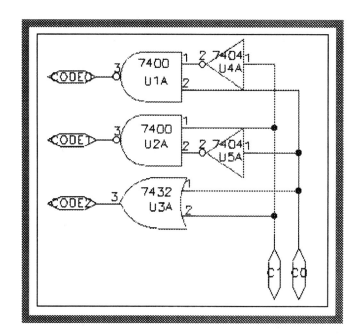

FIGURE 25.4
Block CODE design

5. Finally, we repeat the process to create block COUNT of Figure 25.5.

> Block COUNT counts up by from 0 to 3 from C0 and C1. When a RESET is received, the count resets to zero. When the count reaches three, OPEN is output.

6. Our design complete, we perform a transient analysis from 0 to 10µs and display the curves of Figure 25.6.

　　Is the TWO, ZERO, ONE sequence of U1 the correct OPEN sequence?

　　Yes　　　　No

7. Add CLOCK and RESET signals to your curve set.

　　Is RESET low whenever CLOCK is low (preventing a "clear count" signal from resetting the counter)?

　　Yes　　　　No

PSpice for Windows

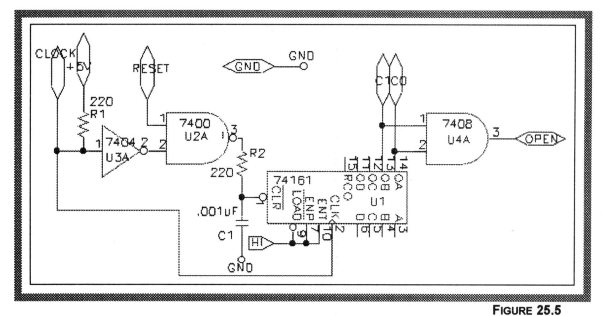

FIGURE 25.5
Block COUNT design

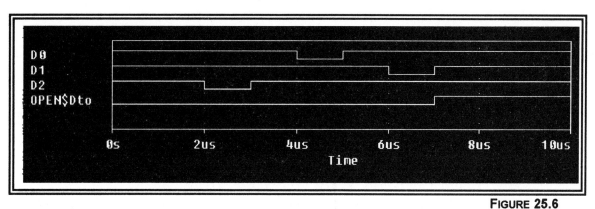

FIGURE 25.6
ALARM output waveforms

Test and modification suggestions

8. Change the input code from the correct TWO, ZERO, ONE sequence to any other sequence and reanalyze the circuit.

    Was OPEN issued?

        Yes   No

9. Add CLOCK and RESET signals to your curve set.

   Is an active-low CLR signal issued to the 74161 counter of block COUNT at the instant the input code varies from the programmed code? Is the CLR signal issued when RESET is high and CLOCK is low?

   Yes   **No**

QUESTIONS & PROBLEMS

1. Show how to reprogram module CODE for a TWO, ZERO, THREE sequence.

2. Why is it necessary to qualify NAND gate U2A or block COUNT with the CLOCK signal?

3. In general terms, how would you modify the overall circuit so a second input of the sequence would return OPEN to zero and SET the alarm?

PSpice for Windows

APPENDIX A
Notes

Schematics

12.1   How do I add a system bus?
12.2   How do I use the digital stimulus editor?

Probe

7.1   How do I create "custom" waveforms out of voltage segments?
11.1   How do I expand (zoom in on) digital waveforms?
11.2   How do I change the size of the digital display?

Filter Synthesis

19.1   What is the meaning of the various preferences options?

PSpice for Windows

APPENDIX B
Probe's Mathematical Operators

Probe Function	Description
( )	Grouping
-	Logical complement
*/	Multiply/divide
+−	Add/subtract
&^|	AND, Exclusive OR, OR
AVGX(x,d)	Average (x to d)
RMS(x)	RMS average
DB(x)	x in dB
MIN(x)	Minimum real part of x
MAX(x)	Maximum real part of x
ABS(x)	Absolute value of x
SGN(x)	+1 if X>0, 0 if x = 0, -1 if x<0
SQRT(x)	$x^{1/2}$
EXP(x)	e^x
LOG(x)	Ln(x)
LOG10(x)	log(x)
M(x)	Magnitude of x
P(x)	Phase of x (degrees)
R(x)	Real part of x
IMG(x)	Imaginary part of x
G(x)	Group delay of x (sec)
PWR(x,y)	$(x)^y$
SIN(x)	sin(x)
COS(x)	cos(x)
TAN(x)	tan(x)
ATAN(x)	$\tan^{-1}(x)$
d(x)	Derivative of x with X-axis
s(x)	Integral of x with X-axis
AVG(x)	Average of x

APPENDIX C
Scale Suffixes

Symbol	Scale	Name
F	10E−15	*femto-*
P	10E−12	*pico-*
N	10E−9	*nano-*
U	10E−6	*micro-*
M	10E−3	*milli-*
K	10E+3	*kilo-*
MEG	10E+6	*mega-*
G	10E+9	*giga-*
T	10E+12	*tera-*

APPENDIX D
Spec Sheets

LM741
OPERATIONAL AMPLIFIER

Parameter	Conditions	Min	Typ	Max	Unit
Input Offset Voltage	TA = 25°C		1.0	5.0	mV
Input Offset Current	TA = 25°C		20	200	nA
Input Bias Current	TA = 25°C		80	500	nA
Input Resistance	TA = 15°C, VS = +- 20V	.3	2.0		Mohms
Large Signal Voltage Gain	TA = 25°C, VS = +- 15V	50	200		V/mV
Output Short Circuit Current	TA = 25°C		25		mA
Common-Mode Rejection Ratio		70	90		DB
Bandwidth	TA = 25°C	.437	1.5		MHz
Slew Rate	TA = 25°C, Unity Gain		.5		V/us
Supply Current	TA = 25°C		1.7	2.8	mA
Power Consumption	TA = 25°C, VS = +- 15V		60	100	mW

PSpice for Windows

TTL FAMILY CHARACTERISTICS

Standard 54/74

Parameter	Test Conditions	Min	Typ	Max	Unit
VIH Input HIGH voltage	Guaranteed input HIGH voltage for all inputs	2.0			V
VIL Input LOW voltage	Guaranteed input LOW voltage for all inputs			.8	V
VCD Input Clamp Diode Voltage	VCC = Min Iin = -12mA		-0.8	-1.5	V
VOL Output LOW voltage	VCC = Min IOL = 16mA			0.4	V
VOH Output HIGH voltage	VCC = Min IOH = -800uA	2.4	3.5		V
IOH Output HIGH current (open collector)	VCC = Max Vout = 5.5V			250	uA
IOZH Output "off" current HIGH (3 state)	VCC = Max Vout = 2.4V VOE = 2.0V			40	uA
IOZL Output "off" current LOW	VCC = Max Vout = .5V VOE = 2.0V			-40	uA
IIH Input HIGH current	VCC = Max Vin = 2.4V			40	uA
IIH Input HIGH current at max input voltage	VCC = Max Vin = 5.5V			1.0	mA
IIL Input LOW current	VCC = Max Vin = 0.4V			-1.6	mA
IOS Output short circuit current	VCC = Max Vout = .0V	-18		-55	mA

Low Power Schottky 54LS/74LS

Parameter	Test Conditions	Min	Typ	Max	Unit
VIH Input HIGH voltage	Guaranteed input HIGH voltage for all inputs	2.0			V
VIL Input LOW voltage	Guaranteed input LOW voltage for all inputs			.8	V
VCD Input Clamp Diode Voltage	VCC = Min Iin = -12mA		-0.65	-1.5	V
VOL Output LOW voltage	VCC = Min IOL = 16mA			0.4	V
VOH Output HIGH voltage	VCC = Min IOH = -800uA	2.7	3.4		V
IOH Output HIGH current (open collector)	VCC = Max Vout = 5.5V			100	uA
IOZH Output "off" current HIGH (3 state)	VCC = Max Vout = 2.4V VOE = 2.0V			20	uA
IOZL Output "off" current LOW	VCC = Max Vout = .5V VOE = 2.0V			-20	uA
IH Input HIGH current	VCC = Max Vin = 2.4V			20	uA
II Input HIGH current at max input voltage	VCC = Max Vin = 5.5V			0.1	mA
IIL Input LOW current	VCC = Max Vin = 0.4V			-0.46	mA
IOS Output short circuit current	VCC = Max Vout = .0V	-18		-100	mA

PSpice for Windows

7442A BCD-to-Decimal Decoder

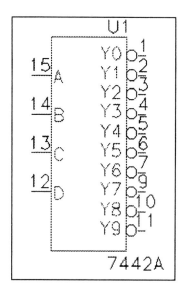

A3	A2	A1	A0	0	1	2	3	4	5	6	7	8	9
L	L	L	L	L	H	H	H	H	H	H	H	H	H
L	L	L	H	H	L	H	H	H	H	H	H	H	H
L	L	H	L	H	H	L	H	H	H	H	H	H	H
L	L	H	H	H	H	H	L	H	H	H	H	H	H
L	H	L	L	H	H	H	H	L	H	H	H	H	H
L	H	L	H	H	H	H	H	H	L	H	H	H	H
L	H	H	L	H	H	H	H	H	H	L	H	H	H
L	H	H	H	H	H	H	H	H	H	H	L	H	H
H	L	L	L	H	H	H	H	H	H	H	H	L	H
H	L	L	H	H	H	H	H	H	H	H	H	H	L

All others   All high

7474 Dual D-type FLIP-FLOP

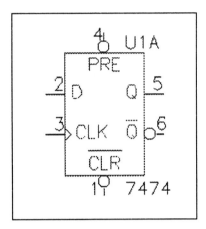

OPERATING MODE	INPUTS				OUTPUTS	
	SD	RD	CP	D	Q	NQ
Asynchronous Set	L	H	X	X	H	L
Asynchronous Reset	H	L	X	X	L	H
Undetermined	L	L	X	X	H	H
Load "1" (Set)	H	H	Rise	H	H	L
Load "0" (Reset)	H	H	Rise	L	L	H

PSpice for Windows

7476 Dual JK FLIP-FLOP

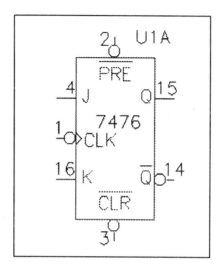

OPERATING MODE	INPUTS					OUTPUTS	
	SD	RD	CP	J	K	Q	NQ
Asynchronous Set	L	H	X	X	X	H	L
Asynchronous Reset	H	L	X	X	X	L	H
Undetermined	L	L	X	X	X	H	H
Toggle	H	H	Pulse	H	H	NQ	Q
Load "1" (Set)	H	H	Pulse	H	L	H	L
Load "0" (Reset)	H	H	Pulse	L	H	L	H
Hold (No change)	H	H	Pulse	L	L	Q	NQ

7485 4-Bit Magnitude Comparator

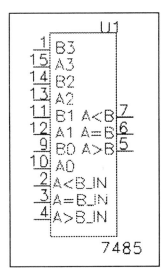

Input signals	A>B	A=B	A<B
A>B	H	L	L
A=B	L	H	L
A<B	L	L	H

7493A 4-Bit Binary Ripple Counter

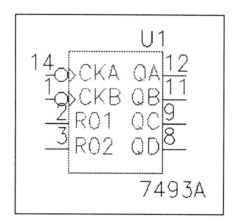

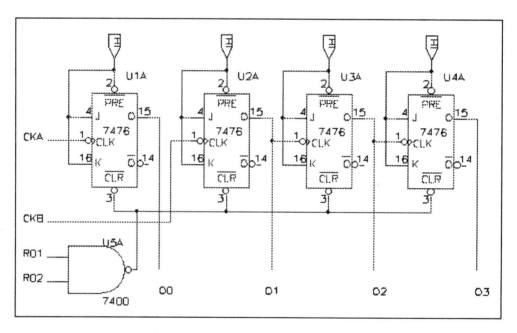

RESET INPUTS		OUTPUTS			
R01	R02	Q0	Q1	Q2	Q3
H	H	L	L	L	L
L	H	Count			
H	L	Count			
L	L	Count			

PSpice for Windows

74147 10-Line-to-4-Line Priority Encoder

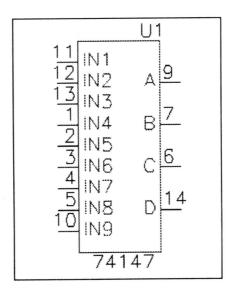

i1	I2	I3	I4	I5	I6	I7	I8	I9	A3	A2	A1	A0
H	H	H	H	H	H	H	H	H	H	H	H	H
X	X	X	X	X	X	X	X	L	L	H	H	L
X	X	X	X	X	X	X	L	H	L	H	H	H
X	X	X	X	X	X	L	H	H	H	L	L	L
X	X	X	X	X	L	H	H	H	H	L	L	H
X	X	X	X	X								
X	X	X	X									
X	X	X										
X	X											
X												
L	H	H	H	H	H	H	H	H				

PSpice for Windows

Index

A

Alarm system 247-252
Ambiguity 142-143, 145
Amplifier
    bandwidth 12
    class B 11-12
    difference 55
    differential 3-10
    instrumentation 89-93
    logarithmic 55
    summing 54, 215
Astable multivibrator 185-187
Asynchronous serial format 233
Asynchronous versus synchronous 165

B

Bandpass filter 215-216
Bandwidth
    general 12
    instrumentation amplifier 91
Big-Ben, clock 241-246
    VCVS 41
Bode plot
    general 30-31
    VCVS 41
Buffer 11-12
Bus, digital 115, 121

C

Calculus 59-69
Closed loop 13-15
CMRR (*see* Common-mode rejection ratio)
Coding, digital 177-184
Common-mode 6
Common-mode rejection ratio (CMRR)
    differential amplifier 7
    instrumentation amplifier 89, 91
    operational amplifier 29
Comparators, digital 177-179, 242-244

Convergence hazard 146-148
Counters, digital
    general 165-176
    Johnson 173
    ripple 167-169
    synchronous 168-170
Critical hazards 157-164
Cumulative ambiguity hazard 148-150
Current-controlled-current-source 50-51
Current-controlled-voltage-source 47-49

D

D-latch
    edge-triggered 135-136
    general 130-131
    level-triggered 133-134
Decoders, digital 180-181
Design methodology, digital 159
Differential amplifier 3-10
    clampers 145, 147-148
    differential mode 5
Differential mode 5
Differentiator 59-69
Displays, digital 105-109

E

Encoders, digital 179-180, 249

F

Filters
    active 82-86, 205-210
    bandpass 215-216
    general 81-87, 195-210
    passive 81, 195-203
    switched capacitor 205, 208-209
Flip-flop
    D-latch 130-131, 133-136
    general 129-140
    JK 132-133

set/reset 111-112
edge/level 129-140

G

Gain-bandwidth product 30
Gates, digital 97-98
Goal function 110

H

Harmonic distortion 12
Hazards, worst-case digital
    ambiguity 145
    convergence 146-148
    critical 157-164
    cumulative ambiguity 148-150
    hold, setup, and width 150-152
    latched 160-162
    persistent 158
    propagation delay 144-145
Hold hazard 150-152

I

ICVS (*see* Current-controlled-voltage-source)
ICIS (*see* Current-controlled-current-source)
Input impedance
    differential amplifier 5, 7
    ICIS 53
    ICVS 52
    op amp 28
    VCIS 43
    VCVS 35, 39
Instrumentation amplifier 89-93
Integrator
    compensated 66
    op amp 59-69
Interface, analog/digital 97-98

J

JK flip-flop 132-133, 137-138
Johnson counter 173

K

L

Latch, D 130-131, 133-136
Latched hazard 160-162

M

Markers, digital 109
Model 17-18, 21
    global vs local 20
Modularization 213-252
Monostable multivibrator 187-189
Moon-landing simulator 67
Multiplexers, digital 182-183
Multiplexing
    FDM versus TDM 223-224

N

Negative feedback 11-15
    closed loop 13-15
    differential amplifier 3
    open loop 11-12
    oscillators 71-79
    shadow rule 36
    VCIS 36-37
    VCVS 33-36

O

Offset voltage 27
Open loop 11-12, 35
Operational amplifier
    common-mode rejection ratio 29
    differentiator 59-69
    filters 81-88
    ideal diode and clamper 43-44
    instrumentation amplifier 89-93
    integrator 59-69
    slew rate, op amp 30
    specifications 25-32
Oscillators
    Colpitts 73-75
    digital 111, 113
    op amp 71-79
    sine wave 72-73
    square wave 76-77
Output impedance
    differential amplifier 5, 7
    ICIS 53
    ICVS 52
    operational amplifier 29
    VCIS 43
    VCVS 35, 40

P

Parity test circuit 138-139
Part 17, 20-21
Propagation delay
   TTL 104-105
      worst-case analysis 141, 144-145
Pulse-code modulation 223-231

Q

R

Rail voltages 27
RS-232C 233-240

S

Sample-and-hold, circuit 220
Serial communication, application 233-240
Setup hazard 150-152
Shadow rule 36, 40
Short circuit
   op amp 28
   TTL 103-104
Slew rate 30
Specifications
   op amp 25-32
   TTL 100-114
Stimulus
   digital 115-127
   editor 122
   file 123-125
Subcircuit 17-34
Summing amplifier 54, 215
Synchronous versus asynchronous 165

T

Timer, 555
   astable multivibrator 185-187
   general 185-191
   monostable multivibrator 187-189
   voltage-controlled oscillator (VCO)
                     189-190
Touch-tone decoding 213-221
Transfer function
   ICVS 49
   ICIS 51
   VCIS 36-37
   VCVS 33, 35
Transistor-transistor-logic
   gates 97-98
   general 97-114
TTL (*see* Transistor-transistor-logic)

U

V

VCIS (*see* Voltage-controlled-current-source)
VCVS (*see* Voltage-controlled-voltage-source)
   transfer function 33-34
Virtual ground 48, 61
Voltage-controlled-current-source 36-37
Voltage-controlled oscillator (VCO)
                     189-190
Voltage-controlled-voltage-source 33-46

W

Width hazard 150-152
Worst case analysis
   ambiguity 142-143, 145
   convergence hazard 146-148
   digital 141-155

X

Y

Z